SCHAUM'S *Easy* OUTLINES

ELECTROMAGNETICS

Other Books in Schaum's Easy Outlines Series Include:

Schaum's Easy Outline: Calculus
Schaum's Easy Outline: College Algebra
Schaum's Easy Outline: College Mathematics
Schaum's Easy Outline: Discrete Mathematics
Schaum's Easy Outline: Elementary Algebra
Schaum's Easy Outline: Geometry
Schaum's Easy Outline: Linear Algebra
Schaum's Easy Outline: Mathematical Handbook of Formulas and Tables
Schaum's Easy Outline: Precalculus
Schaum's Easy Outline: Probability and Statistics
Schaum's Easy Outline: Statistics
Schaum's Easy Outline: Trigonometry
Schaum's Easy Outline: Principles of Accounting
Schaum's Easy Outline: Biology
Schaum's Easy Outline: Biochemistry
Schaum's Easy Outline: College Chemistry
Schaum's Easy Outline: Genetics
Schaum's Easy Outline: Human Anatomy and Physiology
Schaum's Easy Outline: Organic Chemistry
Schaum's Easy Outline: Physics
Schaum's Easy Outline: Programming with C++
Schaum's Easy Outline: Programming with Java
Schaum's Easy Outline: Basic Electricity
Schaum's Easy Outline: Introduction to Psychology
Schaum's Easy Outline: French
Schaum's Easy Outline: German
Schaum's Easy Outline: Spanish
Schaum's Easy Outline: Writing and Grammar

SCHAUM'S *Easy* OUTLINES

ELECTROMAGNETICS

BASED ON SCHAUM'S

Outline of Theory and Problems of Electromagnetics, Second Edition

BY JOSEPH A. EDMINISTER

ABRIDGEMENT EDITOR:

WILLIAM T. SMITH, Ph.D.

SCHAUM'S OUTLINE SERIES

McGRAW-HILL

New York Chicago San Francisco Lisbon London Madrid Mexico City Milan New Delhi San Juan Seoul Singapore Sydney Toronto

JOSEPH A. EDMINISTER is Director of Corporate Relations for the College of Engineering at Cornell University and Professor Emeritus of Electrical Engineering from the University of Akron, where he previously served as professor of electrical engineering, acting department head, assistant dean, and acting dean of engineering. He received bachelor of science in electrical engineering, master of science in engineering, and juris doctorate degrees from the University of Akron and is a registered patent attorney and an attorney in the state of Ohio. He taught electric circuit analysis and electromagnetic theory throughout his academic career.

WILLIAM T. SMITH is an associate professor in the department of electrical engineering at the University of Kentucky, where he has taught since 1990 and has twice won the Outstanding Engineering Professor Award. He earned a B.S. from the University of Kentucky and both M.S. and Ph.D. degrees in electrical engineering from the Virginia Polytechnic Institute and State University. He was an academic visitor at the IBM Austin Research Laboratory and previously worked as a senior engineer in the Government Aerospace Systems Division of Harris Corporation. He is also co-author of several journal articles and conference papers and served as the abridgement editor of *Schaum's Easy Outline: Basic Electricity*.

4 5 6 7 8 9 DOC DOC 0 9 8 7

ISBN 0-07-139879-1

Library of Congress Cataloging-in-Publication Data applied for.

Sponsoring Editor: Barbara Gilson
Production Supervisors: Tama Harris and Clara Stanley
Editing Supervisor: Maureen B. Walker

McGraw-Hill

A Division of The **McGraw-Hill** *Companies*

Contents

Chapter 1
VECTOR ANALYSIS

IN THIS CHAPTER:

- ✔ *Vector Notation*
- ✔ *Vector Algebra*
- ✔ *Coordinate Systems*
- ✔ *Differential Volume; Surface and Line Elements*
- ✔ *Solved Problems*

Vectors are introduced in physics and mathematics courses, primarily in the Cartesian coordinate system. Although cylindrical coordinates may also be found in calculus texts, the spherical coordinate system is seldom presented. All three coordinate systems must be used in electromagnetics. Because the notation, both for the vectors and the coordinate systems, differs from one text to another, a thorough understanding of the notation employed herein is essential for setting up the problems and obtaining solutions.

Vector Notation

In order to distinguish *vectors* (quantities having magnitude and direction) from *scalars* (quantities having magnitude only), the vectors are denoted by boldface symbols. A *unit vector* has a magnitude (or length) of unity and is dimensionless as it indicates only direction. Unit vectors will be denoted by the boldface, lowercase **a**. The unit vectors along the *x, y,* and *z* axes of the Cartesian coordinate system are $\mathbf{a}_x$, $\mathbf{a}_y$, and $\mathbf{a}_z$.

The vector **A** can be written in *component form* as

$$\mathbf{A} = A_x\mathbf{a}_x + A_y\mathbf{a}_y + A_z\mathbf{a}_z$$

In terms of components, the magnitude of the vector **A** is defined as

$$|\mathbf{A}| = A = \sqrt{A_x^2 + A_y^2 + A_z^2}$$

and the unit vector along the direction of **A** is given by

$$\mathbf{a}_A = \frac{\mathbf{A}}{|\mathbf{A}|} = \frac{\mathbf{A}}{A}.$$

Vector Algebra

1. Vectors may be added and subtracted.

$$\begin{aligned}\mathbf{A} \pm \mathbf{B} &= (A_x\mathbf{a}_x + A_y\mathbf{a}_y + A_z\mathbf{a}_z) \pm (B_x\mathbf{a}_x + B_y\mathbf{a}_y + B_z\mathbf{a}_z)\\ &= (A_x \pm B_x)\mathbf{a}_x + (A_y \pm B_y)\mathbf{a}_y + (A_z \pm B_z)\mathbf{a}_z\end{aligned}$$

2. The associative, distributive, and commutative laws apply.

$$\mathbf{A} + (\mathbf{B} + \mathbf{C}) = (\mathbf{A} + \mathbf{B}) + \mathbf{C}$$

$$k(\mathbf{A} + \mathbf{B}) = k\mathbf{A} + k\mathbf{B}, \quad (k_1 + k_2)\mathbf{A} = k_1\mathbf{A} + k_2\mathbf{A}$$

$$\mathbf{A} + \mathbf{B} = \mathbf{B} + \mathbf{A}$$

3. The *dot product* of two vectors is, by definition,

$$\mathbf{A} \bullet \mathbf{B} = AB\cos\theta \text{ (read "}\mathbf{A}\text{ dot }\mathbf{B}\text{")}$$

where θ is the smaller angle between **A** and **B**. In component form, the dot product is equal to

$$\mathbf{A} \bullet \mathbf{B} = A_xB_x + A_yB_y + A_zB_z$$

which gives, in particular,

$$A = |\mathbf{A}| = \sqrt{\mathbf{A} \bullet \mathbf{A}}$$

Note!

Vectors that are *perpendicular* produce a dot product equal to zero due to the included angle being 90°. An important use of this vector property comes in dotting unit vectors:

$\mathbf{a}_x \bullet \mathbf{a}_y = \mathbf{a}_y \bullet \mathbf{a}_z = \mathbf{a}_x \bullet \mathbf{a}_z = 0$

Also note that

$\mathbf{a}_x \bullet \mathbf{a}_x = \mathbf{a}_y \bullet \mathbf{a}_y = \mathbf{a}_z \bullet \mathbf{a}_z = 1$

Example 1.1 The dot product obeys the distributive and scalar multiplication laws

$$\mathbf{A} \bullet (\mathbf{B} + \mathbf{C}) = \mathbf{A} \bullet \mathbf{B} + \mathbf{A} \bullet \mathbf{C}, \quad \mathbf{A} \bullet k\mathbf{B} = k(\mathbf{A} \bullet \mathbf{B})$$

Using the properties of the dot product, prove property 3 from above.

Solution:

$$\begin{aligned}\mathbf{A} \bullet \mathbf{B} &= (A_x\mathbf{a}_x + A_y\mathbf{a}_y + A_z\mathbf{a}_z) \bullet (B_x\mathbf{a}_x + B_y\mathbf{a}_y + B_z\mathbf{a}_z) \\ &= A_xB_x(\mathbf{a}_x \bullet \mathbf{a}_x) + A_yB_y(\mathbf{a}_y \bullet \mathbf{a}_y) + A_zB_z(\mathbf{a}_z \bullet \mathbf{a}_z) \\ &\quad + A_xB_y(\mathbf{a}_x \bullet \mathbf{a}_y) + \cdots + A_zB_y(\mathbf{a}_z \bullet \mathbf{a}_y)\end{aligned}$$

However, $\mathbf{a}_i \bullet \mathbf{a}_i = 1$ ($i = x,y$ or z) and $\mathbf{a}_i \bullet \mathbf{a}_j = 0, i \neq j$. Thus, property 3 above is proven

$$\mathbf{A} \bullet \mathbf{B} = A_xB_x + A_yB_y + A_zB_z$$

4. The *cross product* of two vectors is, by definition,

$$\mathbf{A} \times \mathbf{B} = (AB \sin\theta)\mathbf{a}_n \text{ (read "}\mathbf{A}\text{ cross }\mathbf{B}\text{")}$$

where θ is the smaller angle between **A** and **B**. The unit vector $\mathbf{a}_n$ is normal to the plane determined by **A** and **B** when they are drawn from a common point. There are two normal unit vector directions relative to the plane. The normal unit vector selected is the one in the direction of advance of a right-hand screw when **A** is turned toward **B** (Figure 1-1). This is also referred to as the "right-hand rule" (pull vector **A** into **B** with the fingers of your right hand and the direction of your thumb indicates the proper normal unit vector direction). Because of this direction requirement, the commutative law does not apply to the cross product. Instead,

$$\mathbf{A} \times \mathbf{B} = -\mathbf{B} \times \mathbf{A}$$

Remember

Vectors that are *parallel* produce a cross product equal to zero vector due to the included angle being 0°.

Expanding the cross product in component form,

$$\begin{aligned}\mathbf{A} \times \mathbf{B} &= (A_x\mathbf{a}_x + A_y\mathbf{a}_y + A_z\mathbf{a}_z) \times (B_x\mathbf{a}_x + B_y\mathbf{a}_y + B_z\mathbf{a}_z) \\ &= (A_yB_z - A_zB_y)\mathbf{a}_x + (A_zB_x - A_xB_z)\mathbf{a}_y + (A_xB_y - A_yB_x)\mathbf{a}_z\end{aligned}$$

A mnemonic device used to compute the cross product is to carry out the multiplications as expressed in a determinant

$$\mathbf{A} \times \mathbf{B} = \begin{vmatrix} \mathbf{a}_x & \mathbf{a}_y & \mathbf{a}_z \\ A_x & A_y & A_z \\ B_x & B_y & B_z \end{vmatrix}.$$

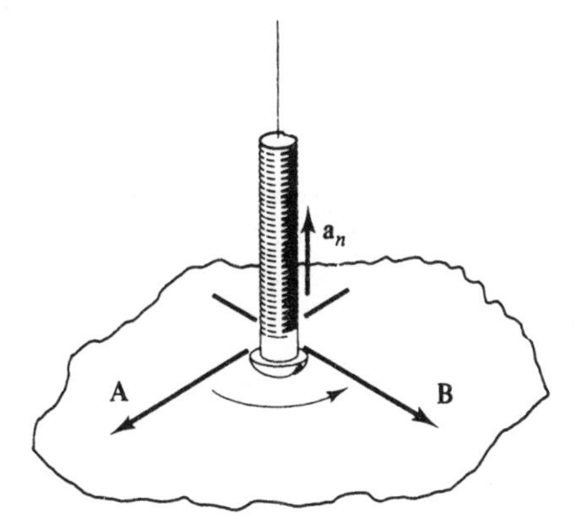

Figure 1-1 Choosing the direction for $\mathbf{a}_n$.

Example 1.2 Given $\mathbf{A} = 2\mathbf{a}_x + 4\mathbf{a}_y - 3\mathbf{a}_z$ and $\mathbf{B} = \mathbf{a}_x - \mathbf{a}_y$, find $\mathbf{A} \bullet \mathbf{B}$ and $\mathbf{A} \times \mathbf{B}$.

Solution:

$$\mathbf{A} \bullet \mathbf{B} = (2)(1) + (4)(-1) + (-3)(0) = -2$$

$$\mathbf{A} \times \mathbf{B} = \begin{vmatrix} \mathbf{a}_x & \mathbf{a}_y & \mathbf{a}_z \\ 2 & 4 & -3 \\ 1 & -1 & 0 \end{vmatrix} = -3\mathbf{a}_x - 3\mathbf{a}_y - 6\mathbf{a}_z$$

Coordinate Systems

A problem that has cylindrical or spherical symmetry could be expressed and solved in the familiar Cartesian coordinate system. However, the solution would fail to show the symmetry and in most cases would be needlessly complex. For example, wire problems use cylindrical coordinates and antenna problems have solutions using spherical coordinates.

A point P is described by three coordinates, in Cartesian (x, y, z), in circular cylindrical (r, ϕ, z), and in spherical (r, θ, ϕ), as shown in Figure 1-2. The same symbol r is used in both cylindrical and spherical coordinates for two different things. In cylindrical coordinates, r is the perpen-

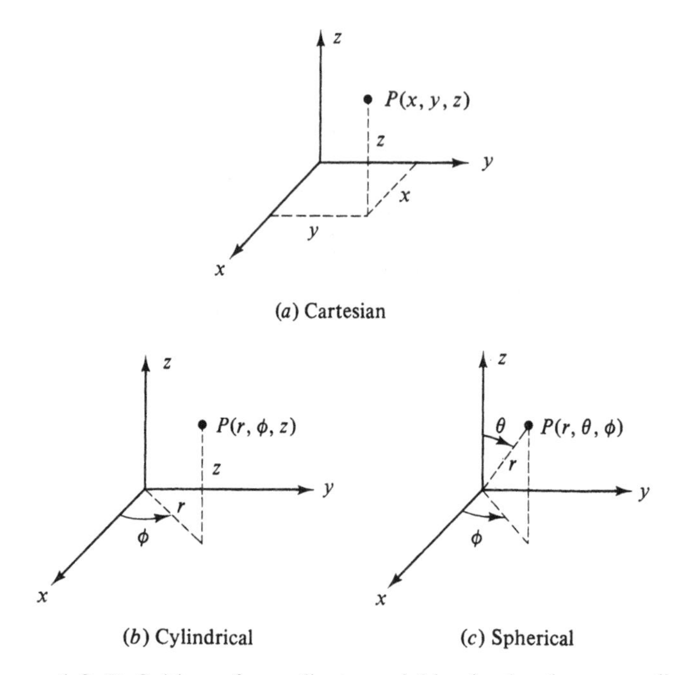

Figure 1-2 Definition of coordinate variables in the three coordinate systems.

dicular distance between the z-axis and the point P. In spherical coordinates, r is the distance from the origin to the point P. It should be clear from the context of the problem which r is intended. The angle ϕ is the same angle in both cylindrical and spherical coordinate systems. The limits are $0 \leq \phi \leq 2\pi$. ϕ is measured in the xy-plane beginning along the x-axis and increasing toward the y-axis. In spherical coordinates, the angle θ is measured between the z-axis and a line from the origin through the point P. The limits are $0 \leq \theta \leq \pi$. The variable z is the same in both Cartesian and cylindrical coordinates.

A point may also be described by the intersection of three orthogonal surfaces as shown in Figure 1-3. In Cartesian coordinates the surfaces are the infinite planes x = const., y = const., and z = const. In cylindrical coordinates, z = const. is the same infinite plane as in Cartesian; ϕ = const. is a half plane with its edge along the z-axis; r = const. is a right

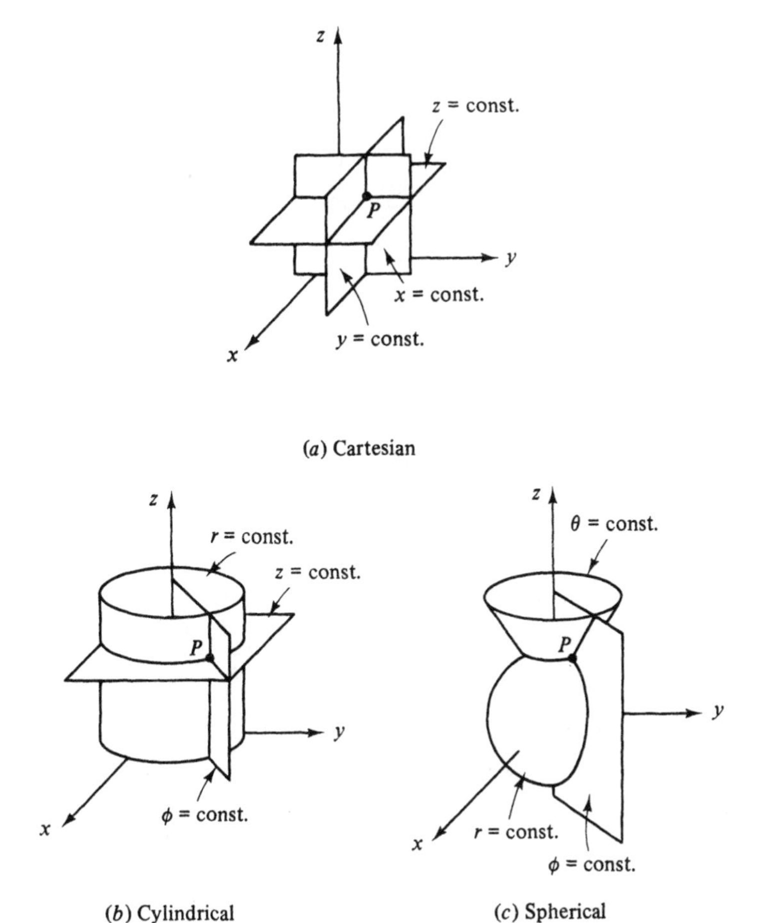

Figure 1-3 Surfaces of constant values for the three coordinate systems.

circular cylinder. These three cylindrical-coordinate surfaces are orthogonal and their intersection locates point P. In spherical coordinates, $\phi =$ const. is the same half plane as in cylindrical; $r =$ const. is a sphere with its center at the origin; $\theta =$ const. is a circular cone whose axis is the z-axis and whose vertex is at the origin. These surfaces are also orthogonal and their intersection once again locates the point P.

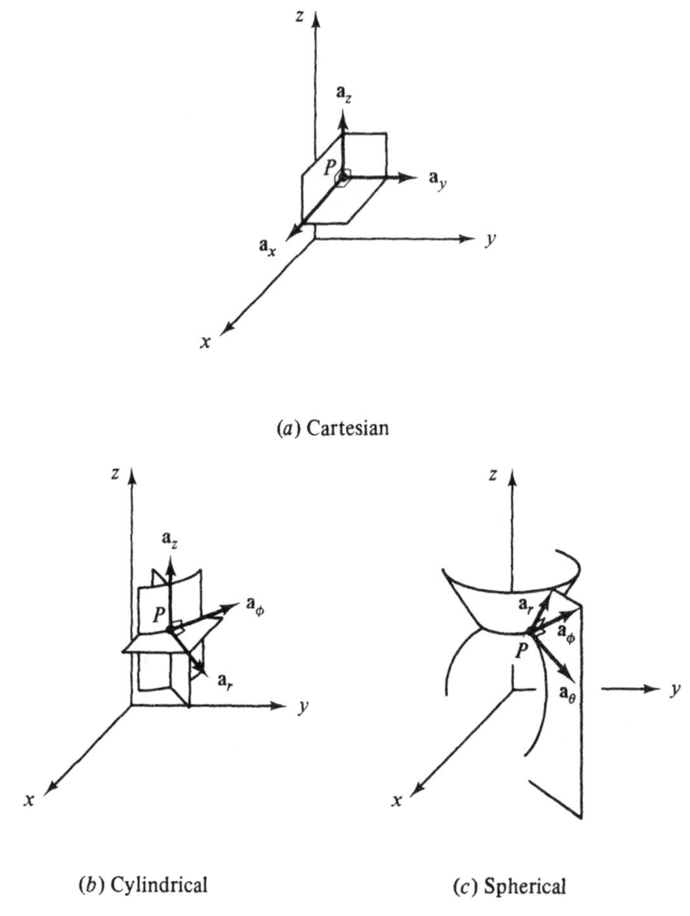

Figure 1-4 Unit vector directions for the three coordinate systems.

Figure 1-4 shows the three unit vectors at point P. In the Cartesian system, the unit vectors have fixed directions, independent of the location of P. This is not true for the other two systems (except for $\mathbf{a}_z$). Each unit vector is normal to its coordinate surface and is in the direction in which the coordinate increases.

Note!

All the systems are right handed:

$$\mathbf{a}_x \times \mathbf{a}_y = \mathbf{a}_z \qquad \mathbf{a}_r \times \mathbf{a}_\phi = \mathbf{a}_z \qquad \mathbf{a}_r \times \mathbf{a}_\theta = \mathbf{a}_\phi$$

The component forms of a vector in the three systems are

$$\mathbf{A} = A_x\mathbf{a}_x + A_y\mathbf{a}_y + A_z\mathbf{a}_z \text{ (Cartesian)}$$

$$\mathbf{A} = A_r\mathbf{a}_r + A_\phi\mathbf{a}_\phi + A_z\mathbf{a}_z \text{ (cylindrical)}$$

$$\mathbf{A} = A_r\mathbf{a}_r + A_\theta\mathbf{a}_\theta + A_\phi\mathbf{a}_\phi \text{ (spherical)}$$

It should be noted that the components $A_x, A_r, A_y, \ldots,$ are not generally constants but more often are functions of the coordinates in that particular system.

Differential Volume, Surface, and Line Elements

There are relatively few problems in electromagnetics that can be solved without some sort of integration—along a curve, over a surface, or throughout a volume. When the coordinates of point P are expanded to $(x + dx, y + dy, z + dz)$ or $(r + dr, \phi + d\phi, z + dz)$ or $(r + dr, \theta + d\theta, \phi + d\phi)$, a differential volume dv is formed. To the first order in infinitesimal quantities, the differential volume is, in all coordinate systems, a box. The value of dv in each system is given in Figure 1-5. The angles θ, ϕ do not have length units as they are angles.

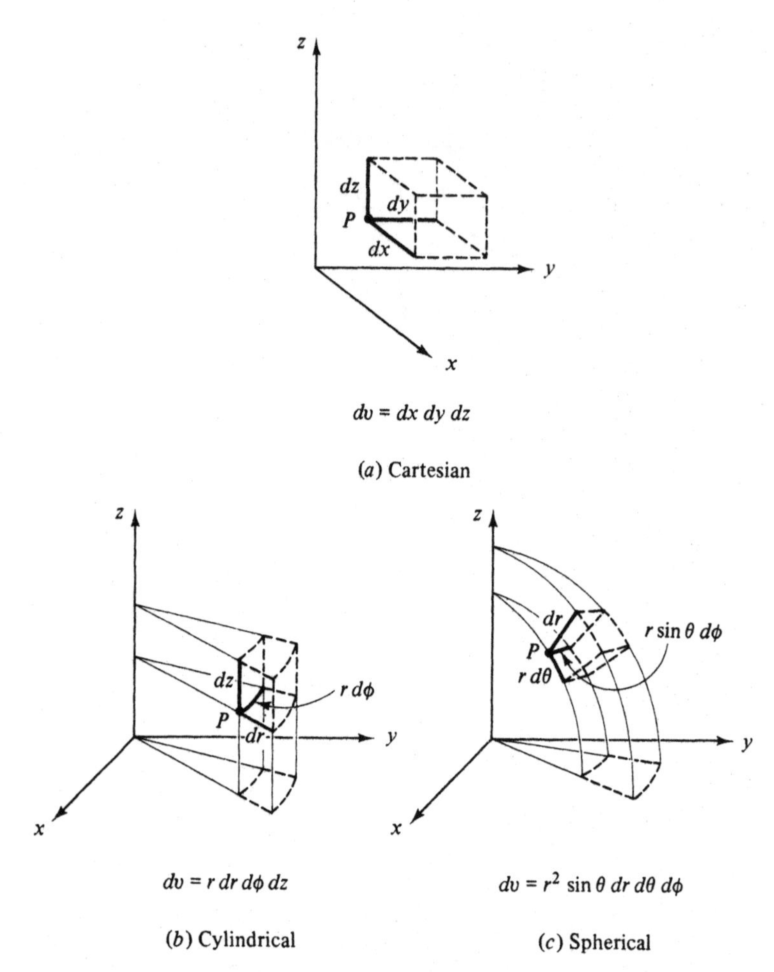

(b) Cylindrical (c) Spherical

Figure 1-5 Differential volumes in the three coordinate systems.

In cylindrical coordinates, the differential length along the circular arc in the ϕ-direction is given by $r\, d\phi$ [see Figure 1-5(*b*)]. In spherical coordinates, the differential lengths along the circular arcs in the θ and ϕ-directions are given by $r\, d\theta$ and $r \sin\theta\, d\phi$, respectively [see Figure 1-5(*c*)].

From Figure 1-5 may also be read the areas of the surface elements that bound the differential volume. For example, in spherical coordinates, the differential surface element perpendicular to $\mathbf{a}_r$ is

$$dS = (r\, d\theta)\, (r \sin\theta\, d\phi) = r^2 \sin\theta\, d\theta\, d\phi$$

The differential line element, dl, is the diagonal through P. Thus

$$dl^2 = dx^2 + dy^2 + dz^2 \text{ (Cartesian)}$$

$$dl^2 = dr^2 + r^2 d\phi^2 + dz^2 \text{ (cylindrical)}$$

$$dl^2 = dr^2 + r^2 d\theta^2 + r^2 \sin^2\theta\, d\phi^2 \text{ (spherical)}$$

Important Things to Remember:

- ✔ A vector is a quantity that possesses both magnitude and direction.
- ✔ The direction of a vector is given by a dimensionless unit vector.
- ✔ Vectors that are *perpendicular* produce a dot product equal to zero due to the included angle being 90°.
- ✔ Vectors that are *parallel* produce a cross product equal to zero vector due to the included angle being 0°.
- ✔ The coordinate variables for the three systems are: Cartesian (x, y, z); cylindrical (r, ϕ, z); spherical (r, θ, ϕ).

Solved Problems

Solved Problem 1.1

Find the vector **C** directed from $M(x_1, y_1, z_1)$ to $N(x_2, y_2, z_2)$. What is its magnitude and directional unit vector?

Solution: The coordinates of M and N are used to write the two position vectors **A** and **B** in Figure 1-6.

$$\mathbf{A} = x_1\mathbf{a}_x + y_1\mathbf{a}_y + z_1\mathbf{a}_z$$

$$\mathbf{B} = x_2\mathbf{a}_x + y_2\mathbf{a}_y + z_2\mathbf{a}_z$$

Then

$$\mathbf{C} = \mathbf{B} - \mathbf{A} = (x_2 - x_1)\mathbf{a}_x + (y_2 - y_1)\mathbf{a}_y + (z_2 - z_1)\mathbf{a}_z$$

The magnitude of **C** is $C = |\mathbf{C}| = \sqrt{(x_2 - x_1)^2 + (y_2 - y_1)^2 + (z_2 - z_1)^2}$.

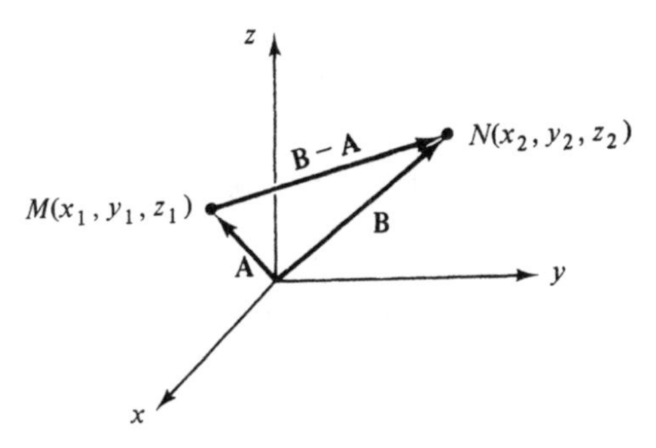

Figure 1-6 Position vectors locating M and N.

The unit vector is $\mathbf{a}_C = \frac{\mathbf{C}}{C} = \frac{(x_2 - x_1)\mathbf{a}_x + (y_2 - y_1)\mathbf{a}_y + (z_2 - z_1)\mathbf{a}_z}{\sqrt{(x_2 - x_1)^2 + (y_2 - y_1)^2 + (z_2 - z_1)^2}}$

Solved Problem 1.2

Find the distance between $(5, 3\pi/2, 0)$ and $(5, \pi/2, 10)$ in cylindrical coordinates.

Solution: First obtain the Cartesian position vectors **A** and **B** (see Figure 1-7).

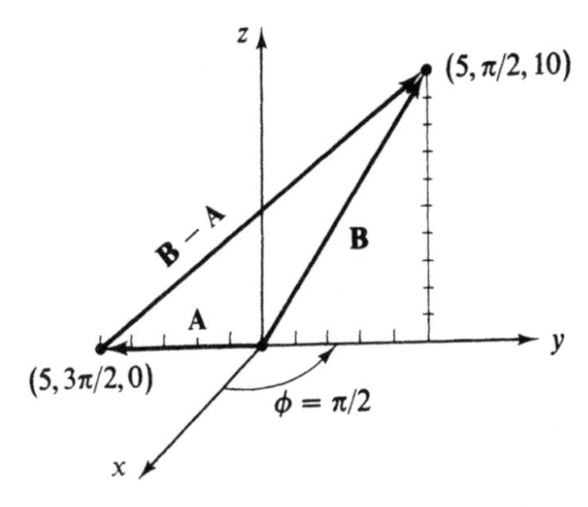

Figure 1-7 Finding the distance between the points.

From the figure,

$$\mathbf{A} = -5\mathbf{a}_y, \qquad \mathbf{B} = 5\mathbf{a}_y + 10\mathbf{a}_z$$

Then, $\mathbf{B} - \mathbf{A} = 10\mathbf{a}_y + 10\mathbf{a}_z$ and the equivalent distance between the points is

$$|\mathbf{B} - \mathbf{A}| = 10\sqrt{2}$$

Solved Problem 1.3

Given $\mathbf{A} = (y - 1)\mathbf{a}_x + 2x\mathbf{a}_y$, find the vector at (2, 2, 1) and its projection on vector $\mathbf{B} = 5\mathbf{a}_x - \mathbf{a}_y + 2\mathbf{a}_z$.

Solution: $\mathbf{A} = (2 - 1)\mathbf{a}_x + 2(2)\mathbf{a}_y = \mathbf{a}_x + 4\mathbf{a}_y$. As indicated in Figure 1-8, the projection of one vector on a second vector is obtained by expressing the unit vector in the direction of the second vector and taking the dot product.

$$\text{Proj. } \mathbf{A} \text{ on } \mathbf{B} = \mathbf{A} \bullet \mathbf{a}_B = \frac{\mathbf{A} \bullet \mathbf{B}}{|\mathbf{B}|}$$

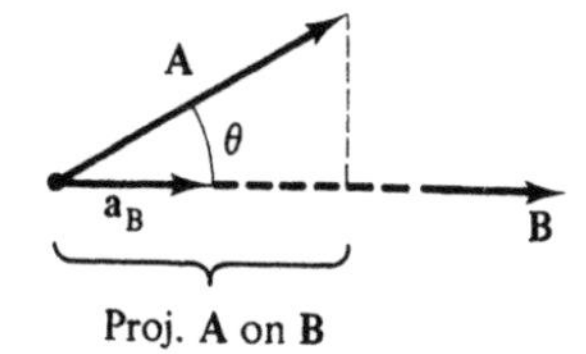

Figure 1-8 Projecting **A** onto **B**.

Thus, at (2, 2, 1),

$$\text{Proj. } \mathbf{A} \text{ onto } \mathbf{B} = \mathbf{A} \bullet \mathbf{a}_B = \frac{\mathbf{A} \bullet \mathbf{B}}{|\mathbf{B}|} = \frac{(1)(5) + (4)(-1) + (0)(2)}{\sqrt{(5)^2 + (-1)^2 + (2)^2}} = \frac{1}{\sqrt{30}}$$

Solved Problem 1.4

Use the spherical coordinate system to find the area of the strip $\alpha \le \theta \le \beta$ on the spherical shell of radius $r = r_0$ (Figure 1-9). What results when $\alpha = 0$ and $\beta = \pi$?

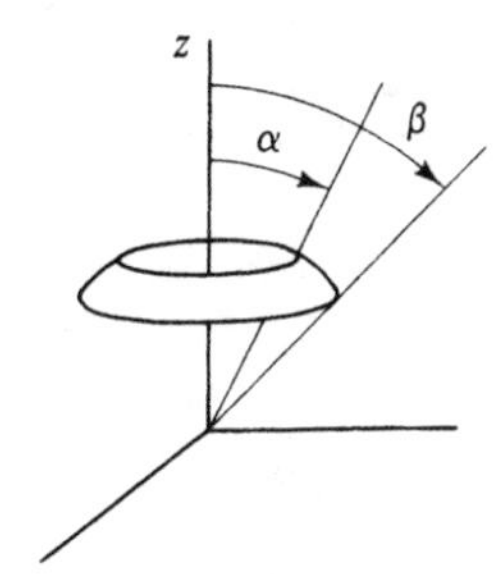

Figure 1-9 Area of a spherical strip.

Solution: The differential surface element is [see Figure 1-5(c)]

$$dS = r_0^2 \sin\theta \, d\theta \, d\phi$$

Then

$$A = \int_0^{2\pi}\int_{\alpha}^{\beta} r_0^2 \sin\theta \, d\theta \, d\phi = 2\pi r_0^2 (\cos\alpha - \cos\beta)$$

When $\alpha = 0$ and $\beta = \pi$, $A = 4\pi r_0^2$, which is the area of the entire sphere.

Chapter 2
STATIC ELECTRIC FIELDS

IN THIS CHAPTER:

- ✔ *Coulomb Forces and Electric Field Intensity*
- ✔ *Electric Flux and Gauss' Law*
- ✔ *Work, Energy, and Potential*
- ✔ *Current and Conductors*
- ✔ *Capacitance*
- ✔ *Solved Problems*

Coulomb Forces and Electric Field Intensity

Coulomb's law was developed from work with small charged bodies and a delicate torsion balance. It describes the force exerted between two electric charges. In vector form, it is stated as

$$\mathbf{F} = \frac{Q_1 Q_2}{4\pi\varepsilon d^2}\mathbf{a}$$

The force is in newtons (N), the distance d in meters (m), and the unit of charge is coulombs (C). The *permittivity* of the medium with units of $C^2/(Nm^2)$ or, equivalently, farads per meter (F/m). In free space or a vacuum, the permittivity is

$$\varepsilon = \varepsilon_0 = 8.854 \times 10^{-12} \approx \frac{10^{-9}}{36\pi} \text{ F/m}$$

For media other than free space, $\varepsilon = \varepsilon_0 \varepsilon_r$, where ε_r is the *relative permittivity* or *dielectric constant*. Free space is to be assumed in all problems and examples unless told otherwise.

You Need to Know 

For point charges of like sign, the Coulomb force is one of repulsion, while for unlike charges the force is attractive.

To incorporate this information, rewrite Coulomb's law as follows:

$$\mathbf{F}_1 = \frac{Q_1 Q_2}{4\pi\varepsilon_0 R_{21}^2} \mathbf{a}_{21} = \frac{Q_1 Q_2}{4\pi\varepsilon_0 R_{21}^3} \mathbf{R}_{21} \quad (1)$$

where $\mathbf{F}_1$ is the force on charge Q_1 due to a second charge Q_2, $\mathbf{a}_{21}$ is the unit vector *directed from Q_2 to Q_1*, and $\mathbf{R}_{21} = R_{21}\mathbf{a}_{21}$ is the position vector from Q_2 to Q_1.

Example 2.1 Find the force on charge Q_1, 20 μC, due to charge Q_2, −300 μC, where Q_1 is at (0,1,2) m and Q_2 is at (2,0,0) m.

Solution: Referring to Figure 2-1, the position vector is

$$\mathbf{R}_{21} = (x_1 - x_2)\mathbf{a}_x + (y_1 - y_2)\mathbf{a}_y + (z_1 - z_2)\mathbf{a}_z$$
$$= (0-2)\mathbf{a}_x + (1-0)\mathbf{a}_y + (2-0)\mathbf{a}_z = -2\mathbf{a}_x + \mathbf{a}_y + 2\mathbf{a}_z$$

$$R_{21} = \sqrt{(-2)^2 + 1^2 + 2^2} = 3$$

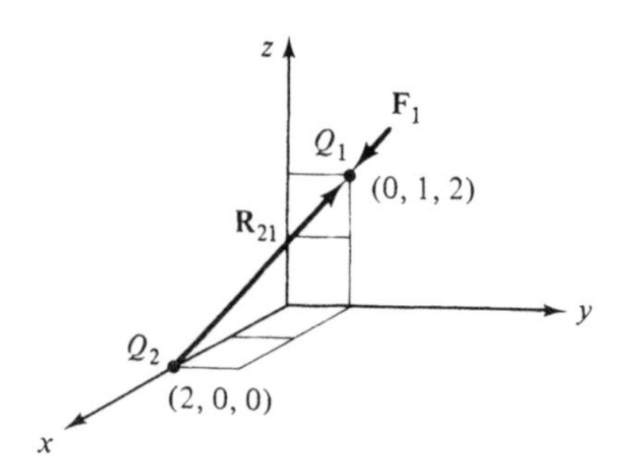

Figure 2-1 Computing the force acting on Q_1.

Using (1), the force is

$$\mathbf{F}_1 = \frac{(20\times10^{-6})(-300\times10^{-6})}{4\pi(10^{-9}/36\pi)(3)^3}(-2\mathbf{a}_x + \mathbf{a}_y + 2\mathbf{a}_z)$$
$$= 4\mathbf{a}_x - 2\mathbf{a}_y - 4\mathbf{a}_z \text{ N}$$

The total force magnitude is 6 N and the direction is such that Q_1 is attracted to Q_2.

The force relationship is bilinear in the charges. Consequently, superposition applies and the force on a charge Q_1 due to n-1 other charges $Q_2 \ldots Q_n$ is the vector sum

$$\mathbf{F}_1 = \frac{Q_1Q_2}{4\pi\varepsilon_0 R_{21}^2}\mathbf{a}_{21} + \frac{Q_1Q_3}{4\pi\varepsilon_0 R_{31}^2}\mathbf{a}_{31} + \ldots = \frac{Q_1}{4\pi\varepsilon_0}\sum_{k=2}^{n}\frac{Q_k}{R_{k1}^2}\mathbf{a}_{k1}$$

If the charge is continuously distributed throughout a region, the above vector sum is replace by a vector integral.

The *electric field intensity* due to a source charge (Q_2 above) is defined to be the force per unit charge on a test charge (Q_1 above):

$$\mathbf{E} = \mathbf{F}_1/Q_1$$

The units of **E** are newtons per coulomb (N/C) or the equivalent, volts per meter (V/m). For a charge Q at the origin of a spherical coordinate system (Figure 2-2), the electric field intensity at a point P is

$$\mathbf{E} = \frac{Q}{4\pi\varepsilon_0 r^2}\mathbf{a}_r \tag{2}$$

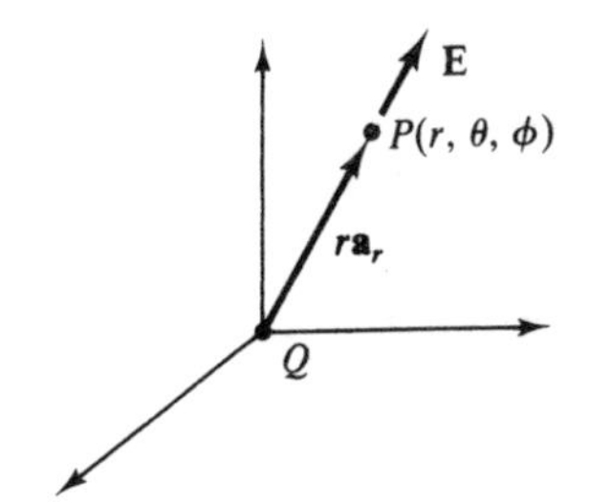

Figure 2-2 Charge Q at the origin.

For Q in an arbitrary Cartesian coordinate system (Figure 2-3),

$$\mathbf{E} = \frac{Q}{4\pi\varepsilon_0 R^2}\mathbf{a}_R \tag{3}$$

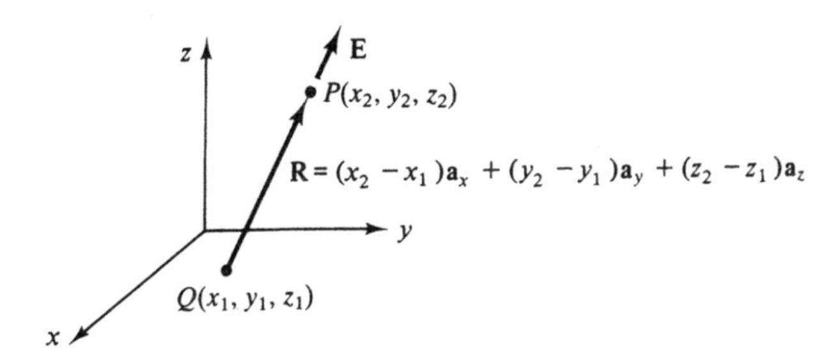

Figure 2-3 Charge Q located at an arbitrary point.

Example 2.2 Find **E** at (0,3,4) m in Cartesian coordinates due to a point charge $Q = 0.5\ \mu\text{C}$ at the origin.

Solution: In this case,

$$\mathbf{R} = (0-0)\mathbf{a}_x + (3-0)\mathbf{a}_y + (4-0)\mathbf{a}_z = 3\mathbf{a}_y + 4\mathbf{a}_z$$

$$R = \sqrt{3^2 + 4^2} = 5$$

$$\mathbf{a}_R = \frac{3\mathbf{a}_y + 4\mathbf{a}_z}{5} = 0.6\mathbf{a}_y + 0.8\mathbf{a}_z$$

Using (3), the electric field intensity is

$$\mathbf{E} = \frac{0.5 \times 10^{-6}}{4\pi(10^{-9}/36\pi)5^2}(0.6\mathbf{a}_y + 0.8\mathbf{a}_z)$$

Thus, $|\mathbf{E}| = 180$ V/m in the direction $0.6\mathbf{a}_y + 0.8\mathbf{a}_z$.

When charge is continuously distributed throughout a specified volume, surface or line, each charge element contributes to the electric field at an external point. For a volume charge density ρ (C/m^3), the elemental charge $dQ = \rho\, dv$ and the differential field at a point P would be (Figure 2-4)

$$d\mathbf{E} = \frac{\rho\, dv}{4\pi\varepsilon_0 R^2}\mathbf{a}_R$$

The total field at the observation point P is obtained by integrating over the volume v

$$\mathbf{E} = \int_v \frac{\rho\, \mathbf{a}_R}{4\pi\varepsilon_0 R^2} dv \qquad (4)$$

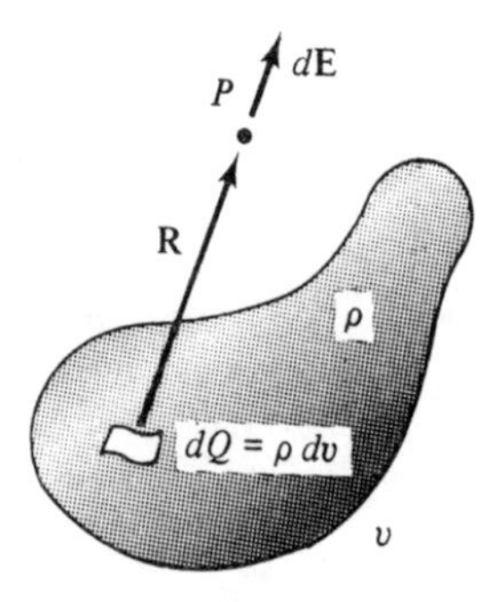

Figure 2-4 **E** due to a volume distribution of charge.

For a surface charge density ρ_s (C/m²), the elemental charge $dQ = \rho_s\, ds$ and the differential field at a point P would be (Figure 2-5)

$$d\mathbf{E} = \frac{\rho_s\, ds}{4\pi\varepsilon_0 R^2}\mathbf{a}_R$$

The total field at the observation point P is obtained by integrating over the surface S

$$\mathbf{E} = \int_S \frac{\rho_s\, \mathbf{a}_R}{4\pi\varepsilon_0 R^2} dS \tag{5}$$

For a linear charge density ρ_ℓ (C/m), the elemental charge $dQ = \rho_\ell d\ell$ and the differential field at a point P would be (Figure 2-6)

$$d\mathbf{E} = \frac{\rho_\ell\, d\ell}{4\pi\varepsilon_0 R^2}\mathbf{a}_R$$

The total field at the observation point P is obtained by integrating over the line or curve L

$$\mathbf{E} = \int_L \frac{\rho_\ell\, \mathbf{a}_R}{4\pi\varepsilon_0 R^2} d\ell \tag{6}$$

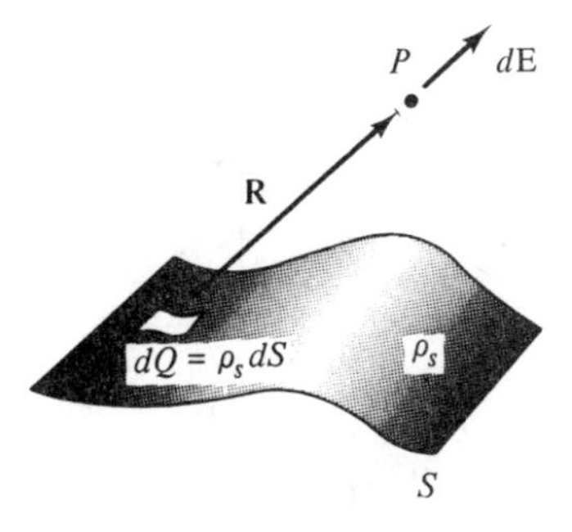

Figure 2-5 E due to a surface distribution of charge.

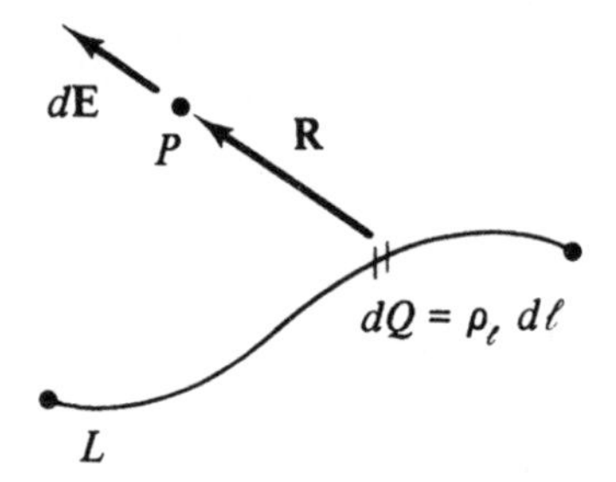

Figure 2-6 E due to a linear distribution of charge.

Three special "standard" charge configurations are the point charge, infinite line charge, and infinite planar surface charge. The **E** for the point charge at the origin is given by equation (2). If the charge density ρ_ℓ is infinite in length and uniformly distributed (constant) along the z-axis, the electric field can be derived using (6) and is (Figure 2-7)

$$\mathbf{E} = \frac{\rho_\ell}{2\pi\varepsilon_0 r}\mathbf{a}_r \quad \text{(cylindrical coodinates)} \tag{7}$$

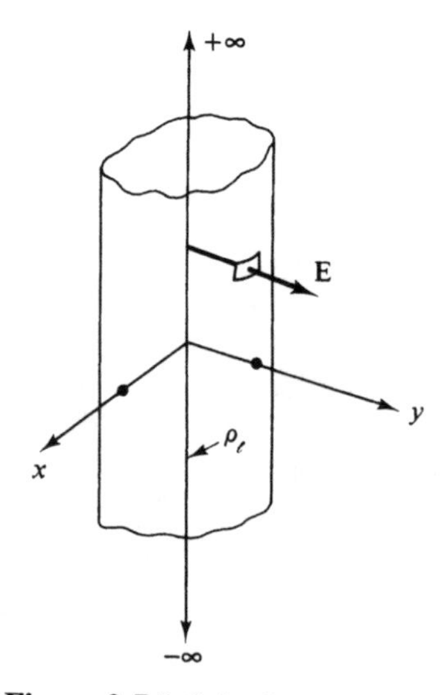

Figure 2-7 Infinite line charge ρ_ℓ.

If the charge is distributed with uniform (constant) density ρ_s over an infinite plane, the field is given by (Figure 2-8)

$$\mathbf{E} = \frac{\rho_s}{2\varepsilon_0}\mathbf{a}_n \tag{8}$$

where $\mathbf{a}_n$ is perpendicular to the surface. The field is of constant magnitude and has mirror symmetry about the planar charge.

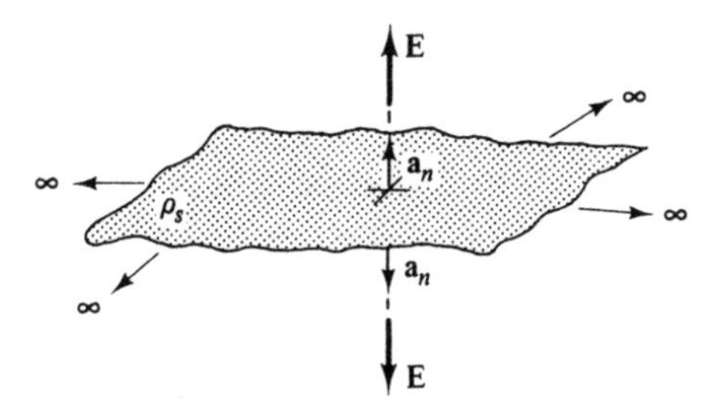

Figure 2-8 Infinite planar surface charge ρ_s.

Example 2.3 Two infinite uniform sheets of charge, each with density ρ_s, are located at $x = \pm 1$ (Figure 2-9). Determine **E** in all regions.

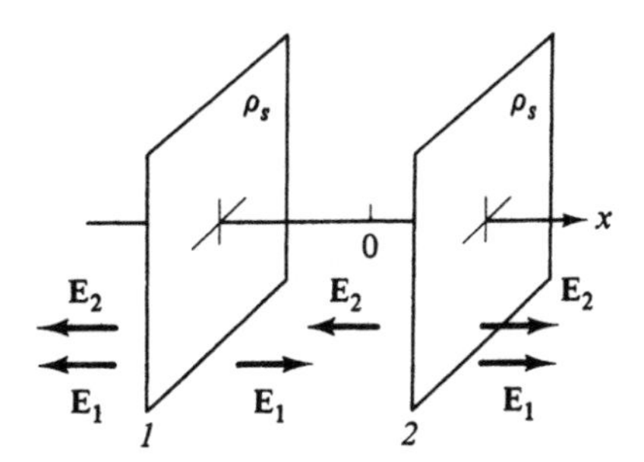

Figure 2-9 Two planar surface charge distributions.

Solution: Only parts of the two sheets of charge are shown in Figure 2-9. Both sheets result in **E** fields that are directed along x, independent of the distance. Using equation (8) and superposition,

$$\mathbf{E} = \begin{cases} -(\rho_s / \varepsilon_0)\mathbf{a}_x & x < -1 \\ 0 & -1 < x < 1 \\ (\rho_s / \varepsilon_0)\mathbf{a}_x & x > 1 \end{cases}$$

Electric Flux and Gauss' Law

Electric flux Ψ is a scalar field and its density **D** is a vector field. By definition, electric flux Ψ emanates from positive charge and terminates on negative charge. In the absence of negative charge, the flux Ψ terminates at infinity. Also by definition, one coulomb of electric charge gives rise to one coulomb of electric flux. Hence,

$$\Psi = Q(\mathrm{C})$$

In Figure 2-10(a), the flux lines leave $+Q$ and terminate on $-Q$. This assumes that the two charges are of equal magnitude. The case of positive charge with no negative charge in the region is illustrated in Figure 2-10(b). Here the flux lines are equally spaced through the angular region surrounding the charge and terminate at infinity.

In the neighborhood of point P, the lines of flux have the direction of a unit vector **a** (Figure 2-11) and if an amount of flux $d\Psi$ crosses the differential surface dS (which is normal to **a**), then the *electric flux density* at P is

$$\mathbf{D} = \frac{d\Psi}{dS}\mathbf{a} \quad (\mathrm{C/m^2})$$

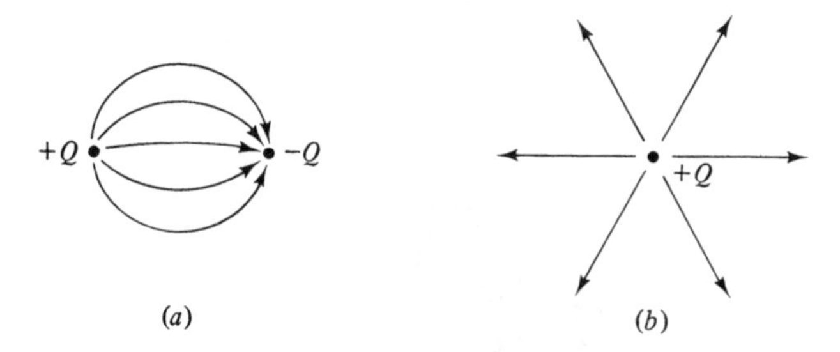

Figure 2-10 Electric flux for point charges.

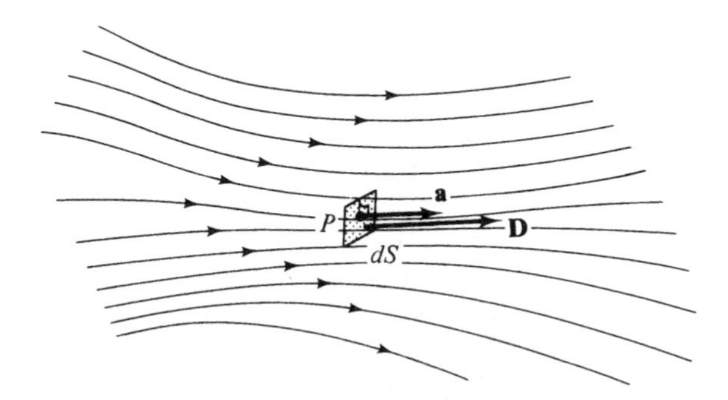

Figure 2-11 Defining the electric flux density **D**.

A volume charge distribution of density ρ (C/m^3) is shown enclosed by surface S in Figure 2-12. Since each coulomb of charge Q has one coulomb of flux, it follows that the net flux crossing the closed surface S is an exact measure of the net charge enclosed. If, at the surface element dS, **D** makes an angle θ with the surface normal unit vector $\mathbf{a}_n$, then the differential flux crossing dS is given by

$$d\Psi = D\,dS\cos\theta = \mathbf{D} \bullet dS\,\mathbf{a}_n = \mathbf{D} \bullet d\mathbf{S}$$

where $d\mathbf{S}$ is the vector surface element. *Gauss' law states that the total flux out of a closed surface is equal to the net charge within the surface.* The *integral form of Gauss' law* is then given by

$$\Psi = \oint_S \mathbf{D} \bullet d\mathbf{S} = Q_{\text{enclosed}} \tag{9}$$

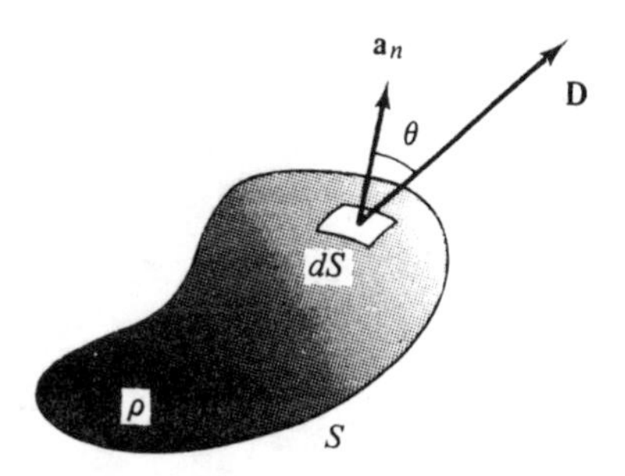

Figure 2-12 Charge density ρ enclosed by surface S.

Consider a point charge Q at the origin (Figure 2-13).

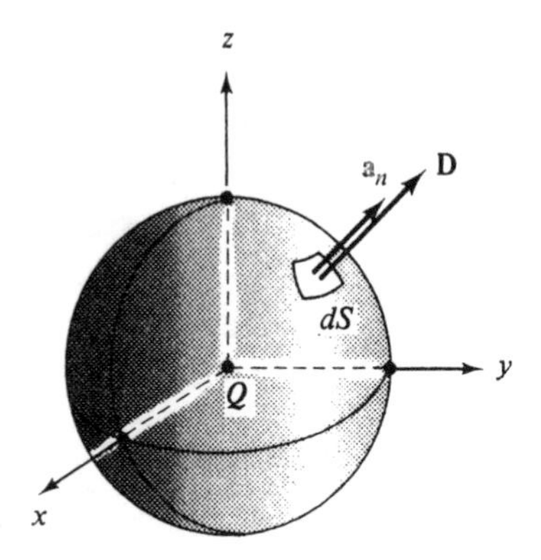

Figure 2-13 Point charge enclosed by a spherical surface.

If this is enclosed by a spherical surface of radius r, then, by symmetry, **D** due to Q is of constant magnitude over the surface and is everywhere normal to the surface. Using Gauss' law (9), the following equality is shown to be

$$Q = \oint_S \mathbf{D} \bullet d\mathbf{S} = D \oint_S dS = D(4\pi r^2)$$

from which $D = Q/4\pi r^2$. Therefore,

$$\mathbf{D} = \frac{Q}{4\pi r^2}\mathbf{a}_r \text{ (spherical coordinates)}$$

Comparing this to (2) gives $\mathbf{D} = \varepsilon_0 \mathbf{E}$. More generally, for any electric field in an *isotropic* medium (medium properties do not vary with field orientation),

$$\mathbf{D} = \varepsilon \mathbf{E}$$

The *divergence of a static electric field* is used to determine when a region has *sources* (*net positive charge*) or *sinks* (*net negative charge*). By definition, the divergence of the electric flux density at a point P is

$$\operatorname{div}\mathbf{D} = \nabla \bullet \mathbf{D} = \lim_{\Delta v \to 0} \frac{\oint_S \mathbf{D} \bullet d\mathbf{S}}{\Delta v} = \lim_{\Delta v \to 0} \frac{Q_{\text{enclosed}}}{\Delta v} = \rho$$

where S is the boundary of Δv.

This is the *point form of Gauss' law.*

$$\nabla \bullet \mathbf{D} = \rho \,(\mathrm{C/m^3}) \tag{10}$$

The point form of Gauss' law gives us a spatial description of the source distribution.

For a general vector $\mathbf{A}$, the definitions for the divergence in the three coordinate systems of interest are:

Cartesian: $$\nabla \bullet \mathbf{A} = \frac{\partial A_x}{\partial x} + \frac{\partial A_y}{\partial y} + \frac{\partial A_z}{\partial z} \tag{11}$$

Cylindrical: $$\nabla \bullet \mathbf{A} = \frac{1}{r}\frac{\partial}{\partial r}(rA_r) + \frac{1}{r}\frac{\partial A_\phi}{\partial \phi} + \frac{\partial A_z}{\partial z} \tag{12}$$

Spherical: $$\nabla \bullet \mathbf{A} = \frac{1}{r^2}\frac{\partial}{\partial r}(r^2 A_r) + \frac{1}{r\sin\theta}\frac{\partial}{\partial \theta}(A_\theta \sin\theta) + \frac{1}{r\sin\theta}\frac{\partial A_\phi}{\partial \phi} \tag{13}$$

Example 2.4 In the region $0 < r < 1$ m, $\mathbf{D} = (-2 \times 10^{-4}/r)\mathbf{a}_r$ (C/m²) and for $r > 1$ m, $\mathbf{D} = (-4 \times 10^{-4}/r^2)\mathbf{a}_r$ (C/m²), in spherical coordinates. Find the charge density in both regions.

Solution: Using (10) and (13), for $0 < r < 1$ m,

$$\rho = \nabla \bullet \mathbf{D} = \frac{1}{r^2}\frac{\partial}{\partial r}\left(r^2 D_r\right) = \frac{1}{r^2}\frac{\partial}{\partial r}\left(-2\times 10^{-4} r\right) = \frac{-2\times 10^{-4}}{r^2}\ (\mathrm{C/m^3})$$

and for $r > 1$ m,

$$\rho = \nabla \bullet \mathbf{D} = \frac{1}{r^2}\frac{\partial}{\partial r}\left(-4\times 10^{-4}\right) = 0$$

The integral and point forms of Gauss' laws are related by the *divergence theorem* given by

$$\Psi = \oint_S \mathbf{D} \bullet d\mathbf{S} = \int_v (\nabla \bullet \mathbf{D})\, dv = Q_{\text{enclosed}}$$

where S is the closed surface boundary of the volume v.

Work, Energy, and Potential

A charge Q experiences a force $\mathbf{F}$ in an electric field $\mathbf{E}$. The force is given by

$$\mathbf{F} = Q\mathbf{E}\ (\mathrm{N})$$

In order to maintain the charge in equilibrium, a force $\mathbf{F}_a = -Q\mathbf{E}$ must be applied in opposition (Figure 2-14).

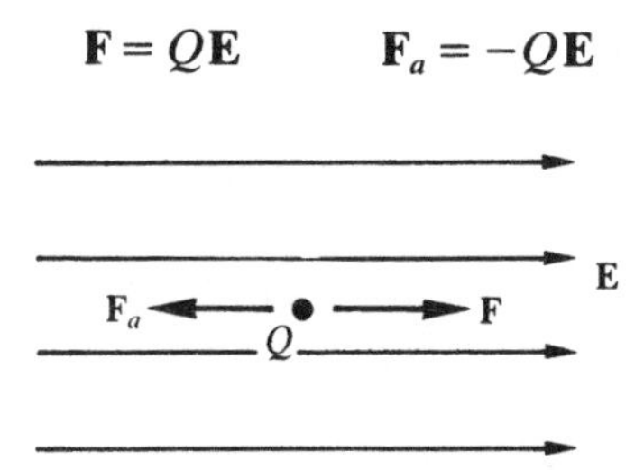

Figure 2-14 Forces acting on charge Q.

> *Work* is defined as force acting over a distance. The units of work are joules (J)

Therefore, a differential amount of work dW is done when the applied force $\mathbf{F}_a$ produces a differential displacement $d\mathbf{l}$ of the charge; i.e., moves the charge through the distance $dl = |d\mathbf{l}|$. Quantitatively,

$$dW = \mathbf{F}_a \bullet d\mathbf{l} = -Q\mathbf{E} \bullet d\mathbf{l} \quad \text{(J)}$$

Note that when Q is positive and $d\mathbf{l}$ is in the direction of $\mathbf{E}$, the work $dW = -Q\,E\,dl < 0$, indicating that *work was done by the electric field.* On the other hand, positive dW indicates *work done against the electric field.*

Component forms of the differential displacement vector are as follows:

Cartesian: $d\mathbf{l} = dx\,\mathbf{a}_x + dy\,\mathbf{a}_y + dz\,\mathbf{a}_z$
Cylindrical: $d\mathbf{l} = dr\,\mathbf{a}_r + r\,d\phi\,\mathbf{a}_\phi + dz\,\mathbf{a}_z$
Spherical: $d\mathbf{l} = dr\,\mathbf{a}_r + r\,d\theta\,\mathbf{a}_\theta + r\sin\theta\,d\phi\,\mathbf{a}_\phi$

Example 2.5 An electrostatic field is given by

$$\mathbf{E} = (x/2 + 2y)\mathbf{a}_x + 2x\mathbf{a}_y \quad \text{(V/m)}$$

Find the work done in moving a point charge $Q = -20\ \mu\text{C}$ (*a*) from the origin to (4,0,0) m, and (*b*) from (4,0,0) m to (4,2,0) m.

Solution:

(a) The first path is along the x-axis, so that $d\mathbf{l} = dx\, \mathbf{a}_x$.

$$dW = -Q\mathbf{E} \bullet d\mathbf{l} = (20 \times 10^{-6})\left(\frac{x}{2} + 2y\right)dx$$

$$W = (20 \times 10^{-6})\int_0^4 \left(\frac{x}{2} + 2y\right)dx\,\Big|_{y=0} = 80\ \mu\text{J}$$

(b) The second path is in the $\mathbf{a}_y$-direction, so that $d\mathbf{l} = dy\, \mathbf{a}_y$.

$$W = (20 \times 10^{-6})\int_0^2 2x\, dy\,\Big|_{x=4} = 320\ \mu\text{J}$$

The work done in moving a point charge from one location A to another location B in the presence of a static electric field is independent of the path taken. Thus, in terms of Figure 2-15,

$$\int_{①} \mathbf{E} \cdot d\mathbf{l} = -\int_{②} \mathbf{E} \cdot d\mathbf{l} \quad \text{or} \quad \oint_{①-②} \mathbf{E} \cdot d\mathbf{l} = 0$$

where the last integral is over the closed contour formed by ① described positively and ② described negatively.

You Need to Know 

For the static electric field, the closed integral is always zero regardless of path and the static electric field is said to be a *conservative field*.

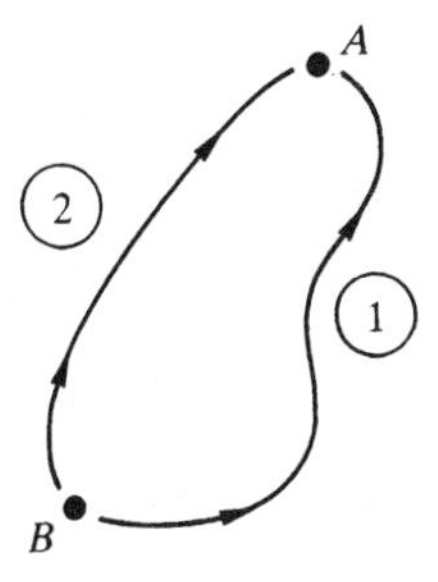

Figure 2-15 Two possible integration paths.

Example 2.6 For the **E** field of Example 2.5, find the work done in moving the same charge from (4,2,0) back to (0,0,0) along a straight-line path.

Solution: The work integral separates into two integrals in x and y.

$$W = (20 \times 10^{-6}) \int_{(4,2,0)}^{(0,0,0)} \left[\left(\frac{x}{2} + 2y \right) \mathbf{a}_x + 2x \mathbf{a}_y \right] \bullet (dx\, \mathbf{a}_x + dy\, \mathbf{a}_y)$$

$$W = (20 \times 10^{-6}) \int_4^0 \left(\frac{x}{2} + 2y \right) dx + (20 \times 10^{-6}) \int_2^0 2x\, dy$$

But, along the path, $y = x/2$. Substituting into the integrals gives

$$W = (20 \times 10^{-6}) \int_4^0 \frac{3}{2} x\, dx + (20 \times 10^{-6}) \int_2^0 4y\, dy = -400\ \mu\text{J}$$

From Example 2.5, $80 + 320 = 400\ \mu$J of work was spent against the field along the outgoing, right-angled path. Exactly this much work was returned by the field along the incoming, straight-line path for a total of zero (conservative field).

The *potential* of point A with respect to point B (denoted as V_{AB}) is defined as the work done in moving a positive charge Q_u from B to A.

$$V_{AB} = \frac{W}{Q_u} = -\int_B^A \mathbf{E} \bullet d\mathbf{l} \quad (\text{J/C or V}) \tag{14}$$

Because the static **E** is a conservative field, $V_{AB} = V_{AC} - V_{CB}$. Therefore, V_{AB} may be considered as the potential difference between points A and B. When V_{AB} is positive, work must be done to move the unit positive charge from B to A and point A is said to be at a *higher potential* than B. Since the electric field of a point charge is completely in the radial direction (2),

$$V_{AB} = -\int_B^A \mathbf{E} \bullet d\mathbf{l} = -\frac{Q}{4\pi\varepsilon_0}\int_{r_B}^{r_A} \frac{dr}{r^2} = \frac{Q}{4\pi\varepsilon_0}\left(\frac{1}{r_A} - \frac{1}{r_B}\right)$$

For a positive charge Q, point A is at higher potential than point B when $r_A < r_B$. If the reference point B is now moved to infinity,

$$V_{A\infty} = \frac{Q}{4\pi\varepsilon_0}\left(\frac{1}{r_A} - \frac{1}{\infty}\right)$$

or

$$V = \frac{Q}{4\pi\varepsilon_0 r}$$

Remember

V is the *absolute potential of Q* referenced to infinity.

If the charge is distributed throughout some finite volume with known charge density ρ (C/m^3), the differential potential at a point P (Figure 2-16) is

$$dV = \frac{dQ}{4\pi\varepsilon_0 R} = \frac{\rho\, dv}{4\pi\varepsilon_0 R}$$

The total potential at point P is found using the integral

$$V = \int_{\text{volume}} \frac{\rho\, dv}{4\pi\varepsilon_0 R}$$

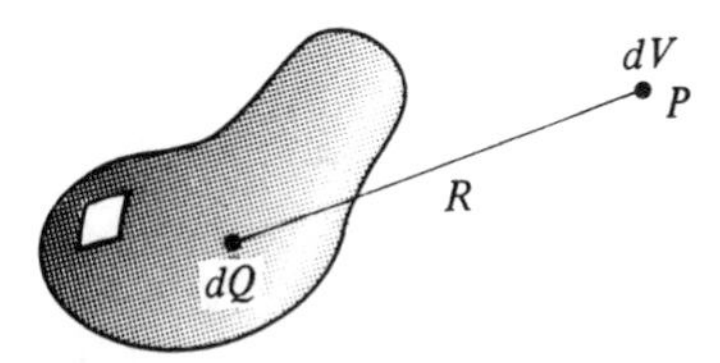

Figure 2-16 Potential of a volume charge density.

Example 2.7 A total charge of 40/3 nano coulomb is uniformly distributed in the form of a circular disk of radius 2 m. Find the potential due to this charge at a point on the axis, 2 m from the disk.

Solution: From Figure 2-17, the cylindrical coordinate system is used to compute the potential.

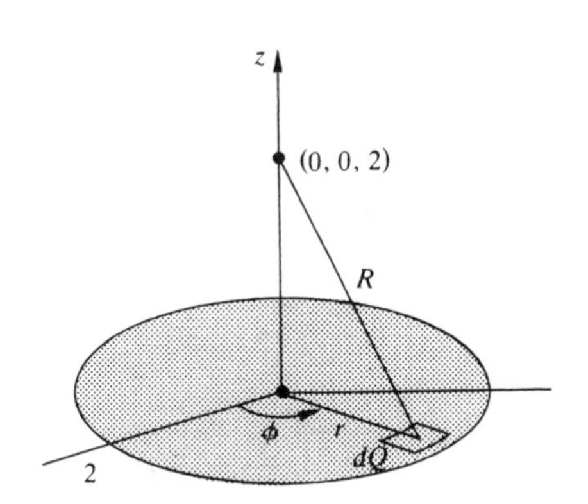

Figure 2-17 Circular disk of surface charge.

For a uniform distribution,

$$\rho_s = \frac{Q}{Area} = \frac{40/3\times10^{-9}}{4\pi} = \frac{10^{-8}}{3\pi} \quad (\text{C/m}^2)$$

The distance R is given by

$$R = \sqrt{4 + r^2} \quad \text{(m)}$$

The potential integral over the surface is

$$V = \int_S \frac{\rho_s \, ds}{4\pi\varepsilon_0 R} = \frac{30}{\pi} \int_0^{2\pi} \int_0^{2} \frac{r \, dr \, d\phi}{\sqrt{4 + r^2}} = 49.7 \text{ V}$$

The electric field and potential are related by the integral (14). A differential relation can also be derived in which the electric field **E** can be found from a known potential V.

The electric field and the potential are also related by

$$\mathbf{E} = -\nabla V$$

where ∇V is the *gradient* of the potential V.

In the coordinate systems of interest, the gradient is defined as

Cartesian: $\nabla V = \frac{\partial V}{\partial x}\mathbf{a}_x + \frac{\partial V}{\partial y}\mathbf{a}_y + \frac{\partial V}{\partial z}\mathbf{a}_z$

Cylindrical: $\nabla V = \frac{\partial V}{\partial r}\mathbf{a}_r + \frac{1}{r}\frac{\partial V}{\partial \phi}\mathbf{a}_\phi + \frac{\partial V}{\partial z}\mathbf{a}_z$

Spherical: $\nabla V = \frac{\partial V}{\partial r}\mathbf{a}_r + \frac{1}{r}\frac{\partial V}{\partial \theta}\mathbf{a}_\theta + \frac{1}{r\sin\theta}\frac{\partial V}{\partial \phi}\mathbf{a}_\phi$

Example 2.8 In spherical coordinates, it was shown that for a charge Q the potential is $V = Q/4\pi\varepsilon_0 r$. Using the spherical gradient,

$$\mathbf{E} = -\nabla V = -\frac{\partial}{\partial r}\left(\frac{Q}{4\pi\varepsilon_0 r}\right)\mathbf{a}_r = \frac{Q}{4\pi\varepsilon_0 r^2}\mathbf{a}_r$$

which is the expression from (2).

Current and Conductors

Electric current is the rate of transport of electric charge past a specified point or across a specified surface. In circuits, the symbol I is generally used for constant currents and i for time-varying currents.

The unit of current is the *ampere.* 1 A = 1 C/s.

Of particular interest is the *conduction current density* **J**. A *conductor* is a material that has a large number of available mobile conduction electrons. The conduction current occurs when an electric field imparts a force upon the mobile electrons and causes an organized flow of the charge throughout the conducting material. The *conductivity* σ of the material has to do with the availability and mobility of the conduction electrons within the material. The conductivity σ has units of Siemans (S).

The relationship between the electric field and the conduction current is given by (Figure 2-18)

$$\mathbf{J} = \sigma \mathbf{E} \quad (\mathrm{A/m^2})$$

This is often referred to as the *point form of Ohm's law.*

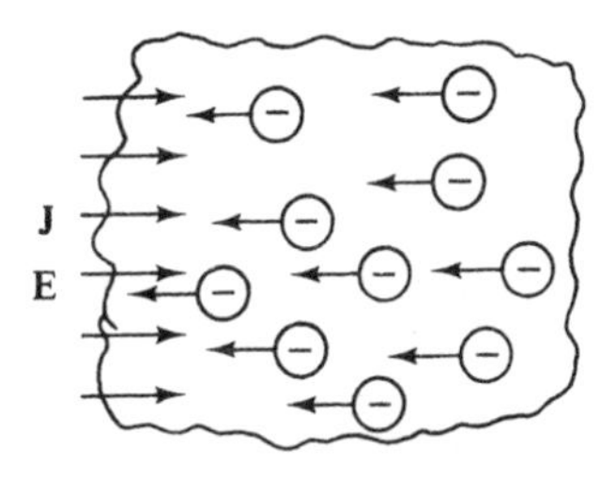

Figure 2-18 Current flow within a conductor.

Where current density **J** crosses a surface S (such as the cross section of a wire), the current I is obtained by integrating the dot product of **J** and the vector differential surface $d\mathbf{S}$ (Figure 2-19)

$$dI = \mathbf{J} \bullet d\mathbf{S} \qquad I = \int_S \mathbf{J} \bullet d\mathbf{S}$$

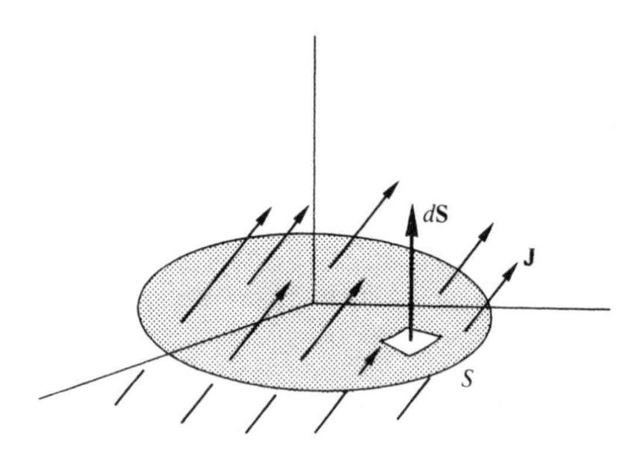

Figure 2-19 **J** flowing through the surface S.

Example 2.9 Find the current in the circular wire shown in Figure 2-20 if the current density is $\mathbf{J} = 15\,(1-e^{-1000r})\,\mathbf{a}_z$ (A/m^2). The radius of the wire is 2 mm.

Solution: A cross section of the wire is chosen for S and the cylindrical coordinate system is used. The differential current is

$$dI = \mathbf{J} \bullet d\mathbf{S} = 15(1-e^{-1000r})\mathbf{a}_z \bullet r\,dr\,d\phi\,\mathbf{a}_z$$

and

$$I = \int_0^{2\pi} \int_0^{0.002} 15(1-e^{-1000r})r\,dr\,d\phi = 1.33\times 10^{-4}\ \text{A} = 0.133\ \text{mA}$$

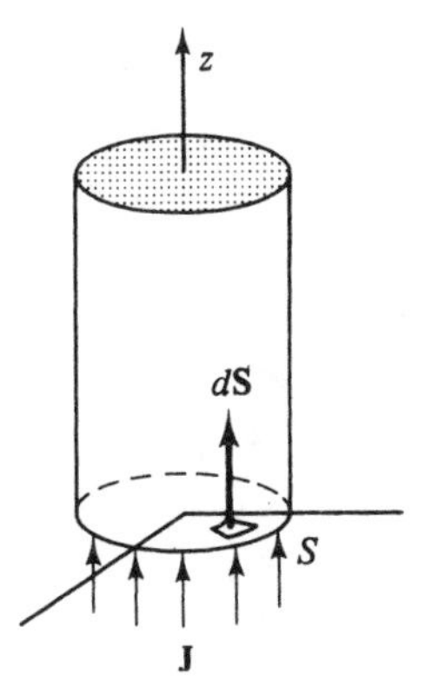

Figure 2-20 Current flowing through a wire.

If a conductor of uniform cross-sectional area A and length ℓ, as shown in Figure 2-21, has a voltage difference V between its ends, then

$$E = \frac{V}{\ell}, \qquad J = \frac{\sigma V}{\ell}$$

assuming that the current is uniformly distributed over the area A. The total current is then

$$I = JA = \frac{\sigma A V}{\ell}$$

Since Ohm's law states that $V = IR$, the *resistance* of the rectangular wire is defined as

$$R = \frac{\ell}{\sigma A} \quad \text{(ohms, } \Omega\text{)}$$

An ohm is related to a Sieman by $1\ \mathrm{S}^{-1} = 1\ \Omega$.

At times, especially at high frequencies, current is confined to the surface of a conductor. For such a surface current density, it is helpful to define the density vector **K** which gives the rate of charge transport per unit length (in A/m). Figure 2-22 shows a total current of I flowing on a cylindrical surface of radius r in the z-direction. In this case, I is uniformly distributed around the periphery of the surface and the surface current density is given by

$$\mathbf{K} = \frac{I}{2\pi r}\mathbf{a}_z$$

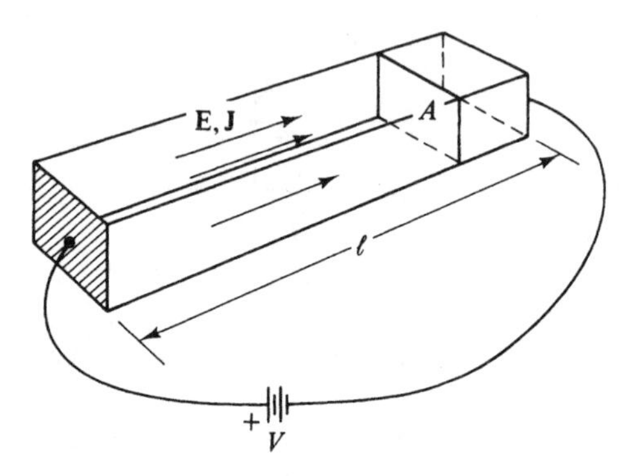

Figure 2-21 Computing the resistance of the conductor.

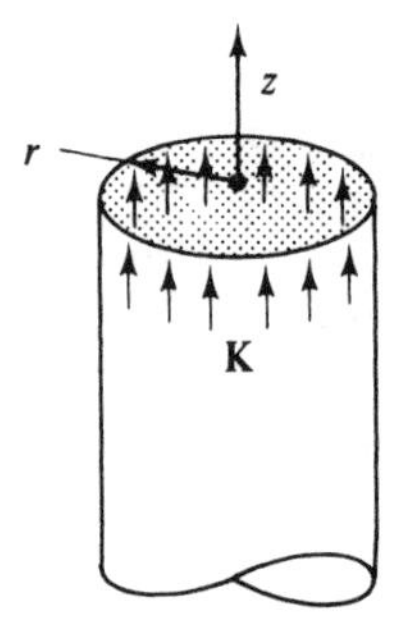

Figure 2-22 Surface current density **K** on a cylinder.

Capacitance

Capacitance is the ability of an object to store electric charge. *Capacitors* are energy-storing circuit elements. To evaluate capacitance, the *boundary conditions* between a conducting material and a dielectric material first need to be defined.

> A *dielectric* is generally thought of as an insulating material.

Under static conditions, all net charge will be on the outer surface of a conductor and both **E** and **D** are zero within the conducting material. Using the conservative property of the static **E** yields (Figure 2-23)

$$\int_1^2 \mathbf{E} \bullet d\mathbf{l} + \int_2^3 \mathbf{E} \bullet d\mathbf{l} + \int_3^4 \mathbf{E} \bullet d\mathbf{l} + \int_4^1 \mathbf{E} \bullet d\mathbf{l} = 0$$

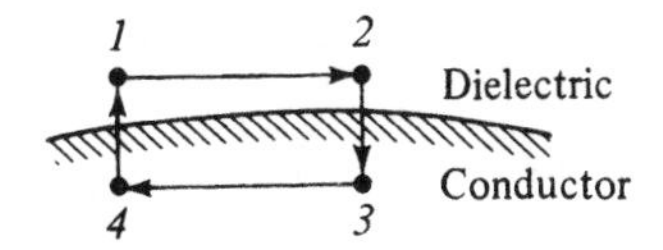

Figure 2-23 Integration path at the boundary between a conductor and a dielectric.

If the path lengths *2* to *3* and *4* to *1* are now permitted to approach zero length, keeping the interface between them, then the second and fourth integrals are zero. The path from *3* to *4* is within the conductor where **E** must be zero. This leaves

$$\int_1^2 \mathbf{E} \bullet d\mathbf{l} = \int_1^2 E_t \, dl = 0$$

where E_t is the tangential component of **E** at the surface of the dielectric.

Note!

The tangential component of **E** and **D** are zero at a conductor-dielectric boundary

$$E_t = D_t = 0$$

To evaluate the conditions on the normal components, a small, closed right-circular cylinder is placed across the interface as shown in Figure 2-24.

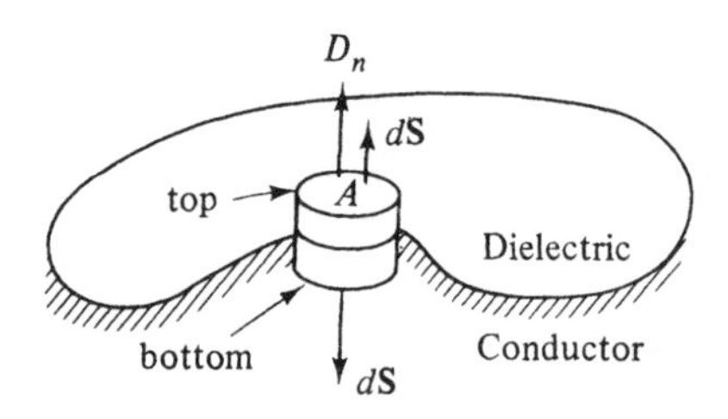

Figure 2-24 Evaluating normal components of fields at a conductor-dielectric boundary.

Gauss' law applied to this surface gives

$$\oint \mathbf{D} \bullet d\mathbf{s} = Q_{enc} = \int_{top} \mathbf{D} \bullet d\mathbf{s} + \int_{bottom} \mathbf{D} \bullet d\mathbf{s} + \int_{side} \mathbf{D} \bullet d\mathbf{s} = \int_{A} \rho_s \, dS$$

The "side" integral goes to zero if the height of the cylinder approaches zero. The "bottom" integral goes to zero as **D** is zero in the conductor. This leaves

$$\int_{top} \mathbf{D} \bullet d\mathbf{s} = \int_{A} D_n dS = \int_{A} \rho_s dS$$

where D_n is the normal component of **D** in the dielectric at the boundary. This can only hold if

$$D_n = \rho_s \qquad \text{and} \qquad E_n = \frac{\rho_s}{\varepsilon}$$

where ε is the permittivity of the dielectric. Thus, normal component of **D** terminates with a surface charge ρ_s at the boundary between a conductor and a dielectric.

Dielectric materials become *polarized* in an electric field which results in an electric flux density **D** that is greater than that under free-space conditions. The polarization effect is caused by an organization of positive/negative *bound* charge pairs within the dielectric called *dipole moments*. This increase in the electric flux density due to polarization is accounted for in isotropic, linear materials through introduction of *permittivity* ε of the material relating the **E** and **D** by

$$\mathbf{D} = \varepsilon \mathbf{E}$$

The permittivity ε is proportional to the permittivity of free-space by

$$\varepsilon = \varepsilon_r \varepsilon_0$$

where ε_r is the *relative permittivity* or *dielectric constant* for the material. For most dielectric materials, $\varepsilon_r > 1$.

Dielectric materials are often used as the insulating materials in *capacitors*. Any two conducting bodies separated by free-space or dielectric material have a *capacitance* between them. A voltage difference V applied results in a $+Q$ on one conductor and a $-Q$ on the other.

The ratio of the absolute value of charge to the absolute value of the voltage difference is defined as the capacitance of the system and is given by

$$C = \frac{Q}{V} \quad \text{(F)}$$

The units of capacitance are farads (F).
1 F = 1 C/V.

The capacitance depends on the geometry of the system and the properties of the dielectric(s) involved.

Example 2.10 Find the capacitance of the parallel plates in Figure 2-25, neglecting fringing (i.e., assume the field is uniformly distributed between the plates and the charge is uniformly distributed on the plates).

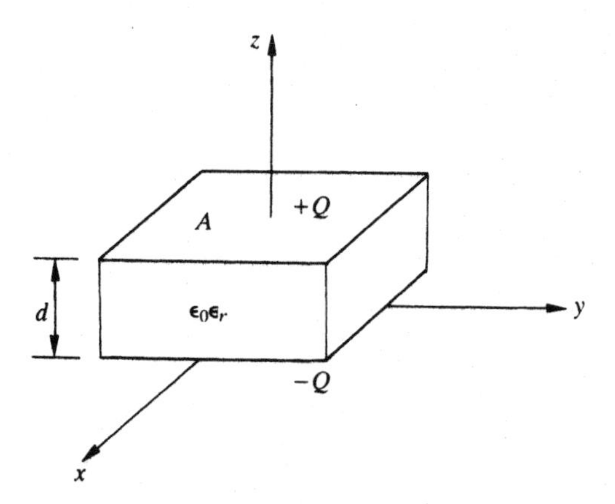

Figure 2-25 Parallel plate capacitor.

Solution: Assume a total charge of $+Q$ on the top plate and $-Q$ on the bottom plate. For uniform distribution of charge on the top plate,

$$\rho_s = \frac{Q}{A}$$

The density on the bottom plate is $-\rho_s$. Between the plates, **D** is uniform and directed from $+\rho_s$ to $-\rho_s$.

$$\mathbf{D} = \frac{Q}{A}(-\mathbf{a}_z) \quad \text{and} \quad \mathbf{E} = \frac{Q}{\varepsilon_0\varepsilon_r A}(-\mathbf{a}_z)$$

The potential of the upper plate with respect to the lower plate is obtained by integrating [using (14)].

$$V = -\int_0^d \left[\frac{Q}{\varepsilon_0\varepsilon_r A}(-\mathbf{a}_z)\right] \bullet (dz\,\mathbf{a}_z) = \frac{Qd}{\varepsilon_0\varepsilon_r A}$$

Then, $C = Q/V = \varepsilon_0\varepsilon_r\ A/d$. Notice that the result only depends on geometry and material permittivity and not the charge or the potential.

When two dielectrics are present in a capacitor with the interface parallel to the **E** and **D**, as shown in Figure 2-26(*a*), the equivalent capacitance can be obtained by treating the arrangement as two parallel capacitors [Figure 2-26(*b*)]

$$C_1 = \frac{\varepsilon_0\varepsilon_{r1}A_1}{d} \qquad C_2 = \frac{\varepsilon_0\varepsilon_{r2}A_2}{d}$$

$$C_{eq} = C_1 + C_2 = \frac{\varepsilon_0}{d}(\varepsilon_{r1}A_1 + \varepsilon_{r2}A_2) \tag{15}$$

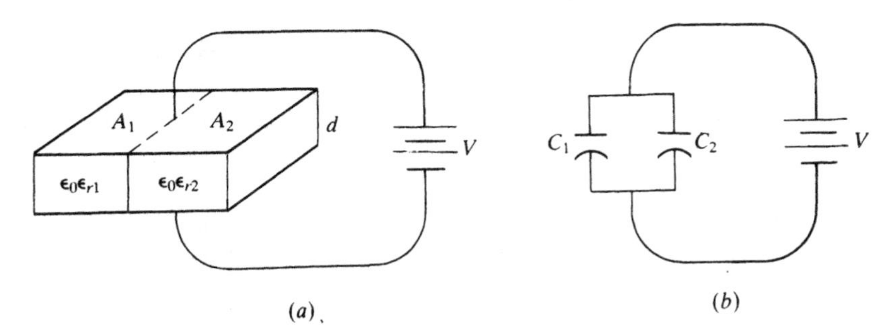

Figure 2-26 Parallel capacitor structure.

When two dielectrics are present such that the interface is normal to **D** and **E**, as shown in Figure 2-27(*a*), the equivalent capacitance can be obtained by treating the arrangement as two capacitors in series [Figure 2-27(*b*)]

$$C_1 = \frac{\varepsilon_0\varepsilon_{r1}A}{d_1} \qquad C_2 = \frac{\varepsilon_0\varepsilon_{r2}A}{d_2}$$

$$\frac{1}{C_{eq}} = \frac{1}{C_1} + \frac{1}{C_2} = \frac{\varepsilon_{r2}d_1 + \varepsilon_{r1}d_2}{\varepsilon_0\varepsilon_{r1}\varepsilon_{r2}A} \tag{16}$$

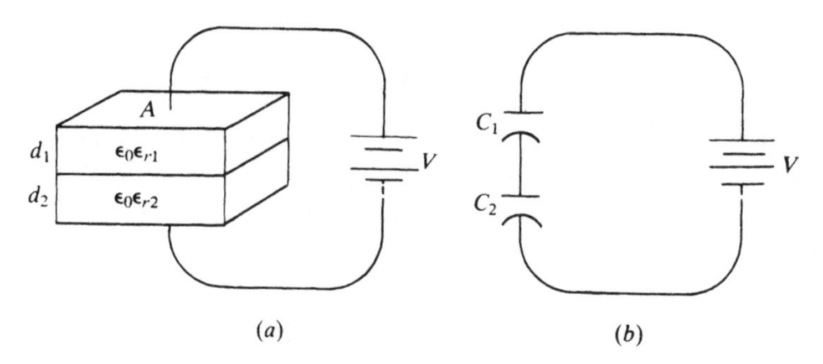

Figure 2-27 Series capacitor structure.

The result can be extended to any number of dielectrics such that the interfaces are all normal to **D** and **E**: *the reciprocal of the equivalent capacitance is the sum of the reciprocals of the individual capacitances.*

Example 2.11 A parallel plate capacitor with area 0.30 m^2 and separation 5.5 mm contains three dielectrics with interfaces normal to **E** and **D**, as follows: $\varepsilon_{r1} = 3.0$, $d_1 = 1.0$ mm; $\varepsilon_{r2} = 4.0$, $d_2 = 2.0$ mm; $\varepsilon_{r3} = 6.0$, $d_3 = 2.5$ mm. Find the capacitance.

Solution: Each dielectric is treated as making up one capacitor in a set of three capacitors in series.

$$C_1 = \frac{\varepsilon_0 \varepsilon_{r1} A}{d_1} = \frac{\varepsilon_0 (3.0)(0.30)}{10^{-3}} = 7.96 \text{ nF}$$

Similarly, $C_2 = 5.31$ nF and $C_3 = 6.37$ nF. Using (16),

$$\frac{1}{C_{eq}} = \frac{1}{7.96\times10^{-9}} + \frac{1}{5.31\times10^{-9}} + \frac{1}{6.37\times10^{-9}} \quad \text{or} \quad C_{eq} = 2.12 \text{ nF}$$

Important Things to Remember:

- ✔ Like charges repel; opposite charges attract.
- ✔ The **E** for a point charge at the origin is radially directed.
- ✔ In an isotropic medium, $\mathbf{D} = \varepsilon\,\mathbf{E}$
- ✔ The **E** and V are related by (14) and $\mathbf{E} = -\nabla V$.
- ✔ Conduction current density is $\mathbf{J} = \sigma\,\mathbf{E}$.
- ✔ For a parallel plate capacitor, the capacitance is given by $C = Q/V = \varepsilon_0 \varepsilon_r A/d$

Solved Problems

Solved Problem 2.1

Refer to Figure 2-28. Find the force on a 100 μC charge at (0,0,3) m if four like charges of 20 μC are located on the x and y axes at ±4 m.

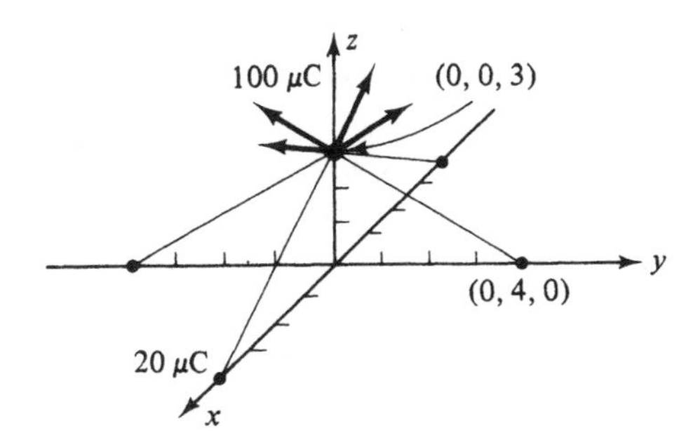

Figure 2-28 Summing forces acting on 100 μC charge.

Solution: Consider the force due to the charge at $y = 4$ m,

$$\frac{(10^{-4})(20\times10^{-6})}{4\pi(10^{-9}/36\pi)(5)^2}\left(\frac{-4\mathbf{a}_y+3\mathbf{a}_z}{5}\right)$$

The y component will be cancelled by the charge at $y = -4$ m. Similarly, the x components due to the other two charges will cancel. Hence,

$$\mathbf{F} = 4\left(\frac{18}{25}\right)\left(\frac{3}{5}\mathbf{a}_z\right) = 1.73\mathbf{a}_z \text{ N}$$

Solved Problem 2.2

Find **E** at (0,0,5) m due to $Q_1 = 0.35$ μC at (0,4,0) m and $Q_2 = -0.55$ μC at (3,0,0) m (see Figure 2-29).

Solution: The distance vectors are

$$\mathbf{R}_1 = -4\mathbf{a}_y + 5\mathbf{a}_z, |\mathbf{R}_1| = \sqrt{41} \text{ m}$$

$$\mathbf{R}_2 = -3\mathbf{a}_x + 5\mathbf{a}_z, |\mathbf{R}_2| = \sqrt{34} \text{ m}$$

$$\mathbf{E}_1 = \frac{0.35\times10^{-6}}{4\text{p}(10^{-9}/36\text{p})(41)}\left(\frac{-4\mathbf{a}_y+5\mathbf{a}_z}{\sqrt{41}}\right)$$
$$= -48.0\mathbf{a}_y + 60.0\mathbf{a}_z \text{ V/m}$$

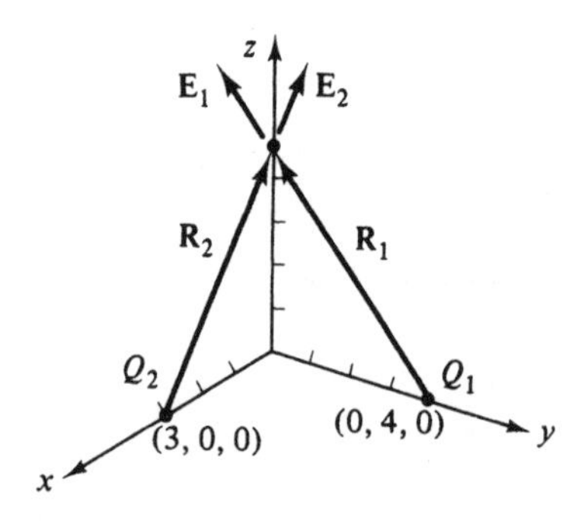

Figure 2-29 Computing **E** at (0,0,5) m.

$$\mathbf{E}_2 = \frac{-0.55\times10^{-6}}{4\pi(10^{-9}/36\pi)(34)}\left(\frac{-3\mathbf{a}_x+5\mathbf{a}_z}{\sqrt{34}}\right)$$
$$= 74.9\mathbf{a}_x - 124.9\mathbf{a}_z \text{ V/m}$$

and the total field at (0,0,5) m is

$$\mathbf{E} = \mathbf{E}_1 + \mathbf{E}_2 = 74.9\mathbf{a}_x - 48\mathbf{a}_y - 64.9\mathbf{a}_z \text{ V/m}$$

Solved Problem 2.3

Find the electric field intensity **E** at $(0,\phi,h)$ due to the uniformly charged disk $\rho_s = const.$, $r \le a$, $z = 0$ (see Figure 2-30).

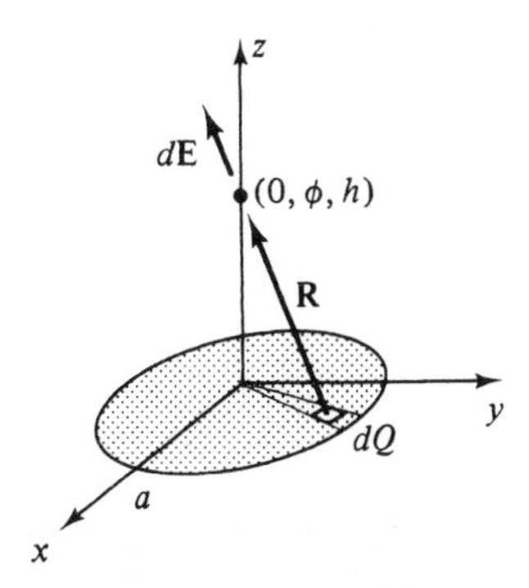

Figure 2-30 E at $(0, \phi, h)$ m due to uniformly charged disk.

The elemental **E** due to an element of ρ_s is

$$d\mathbf{E} = \frac{\rho_s r\,dr\,d\phi}{4\pi\varepsilon_0(r^2+h^2)}\left(\frac{-r\mathbf{a}_r + h\mathbf{a}_z}{\sqrt{r^2+h^2}}\right)$$

The radial components cancel due to symmetry. Therefore,

$$\mathbf{E} = \frac{\rho_s h}{4\pi\varepsilon_0}\int_0^{2\pi}\int_0^{a}\frac{r\,dr\,d\phi}{(r^2+h^2)^{3/2}}\mathbf{a}_z$$
$$= \frac{\rho_s h}{2\varepsilon_0}\left(\frac{-1}{\sqrt{a^2+h^2}}+\frac{1}{h}\right)\mathbf{a}_z \text{ V/m}$$

Note that as $a \to \infty, \mathbf{E} \to (\rho_s/2\varepsilon_0)\mathbf{a}_z$ V/m, the field due to a uniform sheet.

Solved Problem 2.4

The volume in cylindrical coordinates between $r = 2$ m and $r = 4$ m contains a uniform density ρ (C/m^3) (Figure 2-31). Use Gauss' law to find **D** in all regions.

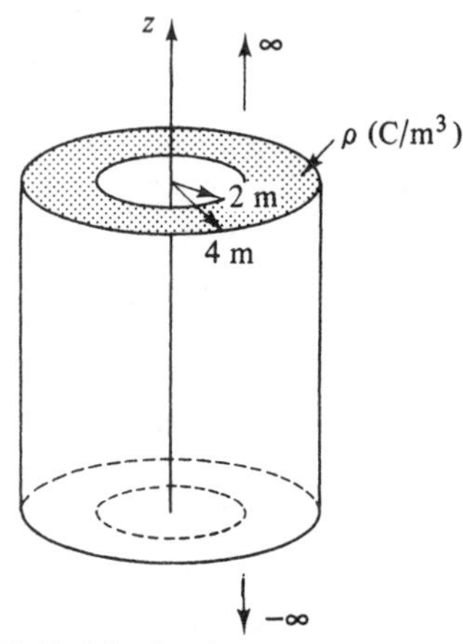

Figure 2-31 Cylindrical volume charge density ρ (C/m^3).

Solution: Since ρ is not a function of ϕ or z, the flux Ψ is completely radial. It is also true that for constant r, the flux density **D** must be of constant magnitude. Then a proper special Gaussian surface is a closed right circular cylinder. The surface integrals from (9) over the plane ends are zero as flux only exits the curved cylinder through the side. For a length L,

$$\oint_S \mathbf{D} \bullet d\mathbf{S} = D(2\pi rL) = Q_{enc}$$

From Figure 2-31, for $0 < r < 2$ m,

$$Q_{enc} = 0 = D(2\pi rL)$$
$$\Rightarrow \mathbf{D} = 0$$

For $2 \le r \le 4$ m,

$$Q_{enc} = \pi\rho L(r^2 - 4) = D(2\pi rL)$$
$$\Rightarrow \mathbf{D} = \frac{\rho}{2r}(r^2 - 4)\,\mathbf{a}_r \text{ C/m}^2$$

For $r \geq 4$ m,

$$Q_{enc} = \pi\rho L(16-4) = D(2\pi rL)$$

$$\Rightarrow \mathbf{D} = \frac{6\rho}{r}\mathbf{a}_r \ \mathrm{C/m^2}$$

Solved Problem 2.5

Find the capacitance of a coaxial capacitor of length L, where the inner conductor has radius a and the outer has radius b (Figure 2-32).

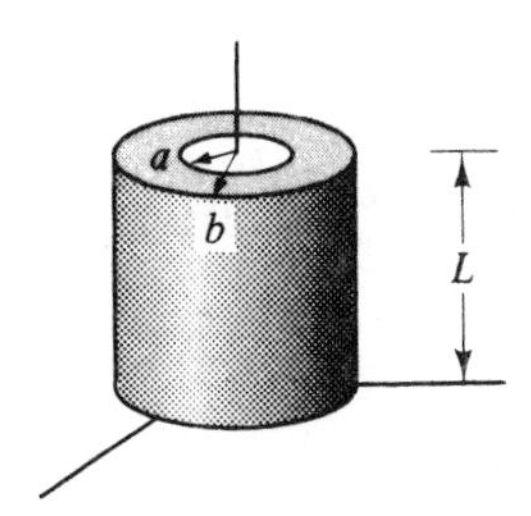

Figure 2-32 Coaxial capacitor.

Solution: With fringing neglected, Gauss' law requires that $\mathbf{D} \propto 1/r$ between the conductors. At $r = a$, $D = \rho_s$ (due to conductor/dielectric boundary conditions), where ρ_s is the (assumed positive) surface charge density on the inner conductor. Therefore,

$$\mathbf{D} = \rho_s \frac{a}{r}\mathbf{a}_r \ \mathrm{C/m^2}, \qquad \mathbf{E} = \frac{\rho_s a}{\varepsilon_0 \varepsilon_r r}\mathbf{a}_r \ \mathrm{V/m}$$

and the voltage difference between the conductors is

$$V_{ab} = -\int_b^a \frac{\rho_s a}{\varepsilon_0 \varepsilon_r r} dr = \frac{\rho_s a}{\varepsilon_0 \varepsilon_r} \ln\frac{b}{a} \ \mathrm{V}$$

The total charge on the inner conductor is $Q = \rho_s (2\pi aL)$ and so

$$C = \frac{Q}{V} = \frac{2\pi\varepsilon_0\varepsilon_r L}{\ln(b/a)} \ \mathrm{F}$$

Chapter 3
STATIC MAGNETIC FIELDS

IN THIS CHAPTER:

- ✔ *Biot-Savart Law*
- ✔ *Ampere's Law*
- ✔ *Magnetic Flux Density and Gauss' Law*
- ✔ *Inductance*
- ✔ *Solved Problems*

A static magnetic field can originate from either a constant current or a permanent magnet. This chapter deals with the fields of constant currents.

Biot-Savart Law

A differential *magnetic field intensity*, $d\mathbf{H}$, results from a differential current element $I\,d\mathbf{l}$. The field varies inversely with the distance squared, is independent of the surrounding medium, and has a direction given by the cross product of $I\,d\mathbf{l}$ and $\mathbf{a}_R$. This relationship is known as the *Biot-Savart law* (Figure 3-1):

$$d\mathbf{H} = \frac{I\,d\mathbf{l} \times \mathbf{a}_R}{4\pi R^2} \quad (\mathrm{A/m}) \qquad (1)$$

where $\mathbf{a}_R$ is the unit vector in the direction of **R**. The direction of **R** must be from the current element to the point at which $d\mathbf{H}$ is to be determined.

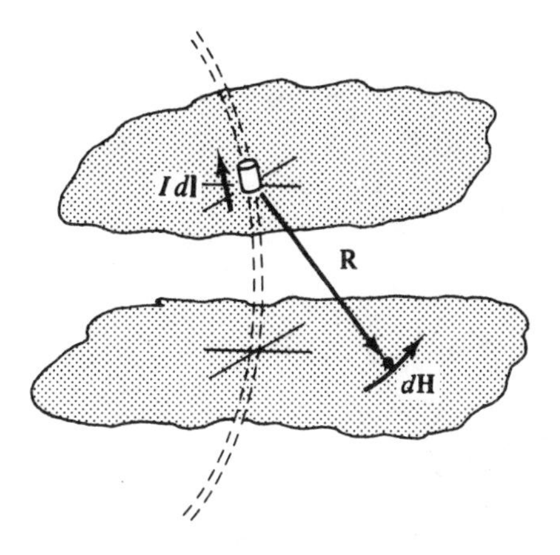

Figure 3-1 Current element giving rise to $d\mathbf{H}$.

Current elements have no separate existence. All elements making up the complete current filament contribute to **H** and must be included. The summation leads to the integral form of the *Biot-Savart law*:

$$\mathbf{H} = \oint \frac{I\,d\mathbf{l} \times \mathbf{a}_R}{4\pi R^2} \quad (\mathrm{A/m})$$

The integral must be performed over a closed current path to ensure that all current elements are included (the path may close at infinity).

Example 3.1 An infinitely long, straight current filament I along the z-axis in cylindrical coordinates is shown in Figure 3-2. Find **H**.

Solution: A point in the $z = 0$ plane is selected with no loss in generality.

$$\mathbf{R} = r\mathbf{a}_r - z\mathbf{a}_z, \quad R = \sqrt{r^2 + z^2}, \quad \mathbf{a}_R = \frac{r\mathbf{a}_r - z\mathbf{a}_z}{\sqrt{r^2 + z^2}}$$

In differential form, using (1),

$$d\mathbf{H} = \frac{I\,dz\,\mathbf{a}_z \times (r\,\mathbf{a}_r - z\,\mathbf{a}_z)}{4\pi(r^2 + z^2)^{3/2}}$$
$$= \frac{I\,r\,dz\,\mathbf{a}_\phi}{4\pi(r^2 + z^2)^{3/2}}$$

The variable of integration is z. Since $\mathbf{a}_\phi$ does not change with z, it may be removed from the integrand before integrating.

$$\mathbf{H} = \left[\int_{-\infty}^{\infty} \frac{I\,r\,dz}{4\pi(r^2 + z^2)^{3/2}}\right]\mathbf{a}_\phi = \frac{I}{2\pi r}\mathbf{a}_\phi$$

This important result shows that $\mathbf{H}$ is inversely proportional to the radial distance.

The direction is seen to be in agreement with the "*right-hand rule*" whereby the fingers of the right hand point in the direction of the field when the conductor is grasped such that the right thumb points in the direction of the current.

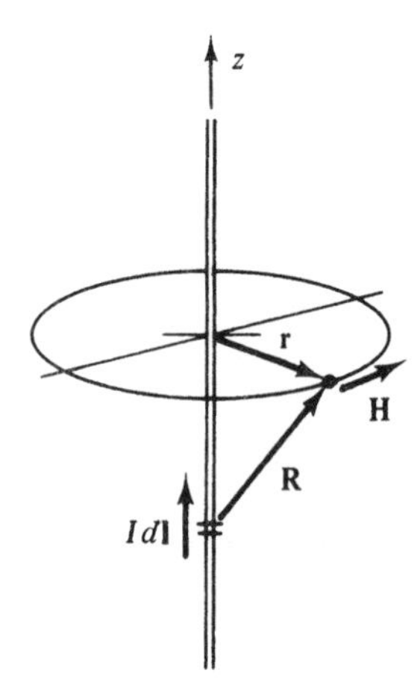

Figure 3-2 Infinite current filament I along the z-axis.

Ampere's Law

The line integral of the tangential component of the magnetic field strength around a closed path is equal to the current enclosed by the path:

$$\oint \mathbf{H} \bullet d\mathbf{l} = I_{enc} \qquad (2)$$

This is the integral form of *Ampere's law.*

At first glance one might think that the law is used to determine the current I by an integration. Instead, the current is usually known and the law provides a method of finding **H**. This is quite similar to the use of Gauss' law to find **D** given the charge distribution.

In order to utilize Ampere's law to determine **H**, there must be a considerable degree of symmetry in the problem. Two conditions must be met:

1. At each point of the close path **H** is either tangential or normal to the path.
2. H has the same value at all points of the path where **H** is tangential.

The Biot-Savart law can be used to aid in selecting a path that meets the above conditions. In most cases a proper path will be evident.

Example 3.2 Use Ampere's law to obtain **H** due to an infinitely long, straight current filament I.

Solution: The Biot-Savart law shows that at each point of the circle in Figure 3-2 **H** is tangential and of the same magnitude. Then,

$$\oint \mathbf{H} \bullet d\mathbf{l} = H(2\pi r) = I$$

Solving,

$$\mathbf{H} = \frac{I}{2\pi r}\mathbf{a}_\phi$$

A differential form of Ampere's law can be derived which also relates the static magnetic field **H** to constant electric current. Before defining the differential form, the *curl of a vector* must be introduced.

The curl of a general vector field **A** is another vector field. Point P in Figure 3-3 lies in a plane area ΔS bounded by a closed curve C. In the integration that defines the curl, C is traversed such that the enclosed area is on the left. The unit normal $\mathbf{a}_n$, determined by a right-hand rule (fingers of right hand curl in direction of integration along the contour, thumb points in direction of the normal), is as shown in the figure. Then the *component* of the curl of **A** in the direction $\mathbf{a}_n$ is defined as

$$(\text{curl}\,\mathbf{A})\bullet\mathbf{a}_n = \nabla\times\mathbf{A}\bullet\mathbf{a}_n \equiv \lim_{\Delta S\to 0}\frac{\oint \mathbf{A}\bullet d\mathbf{l}}{\Delta S}$$

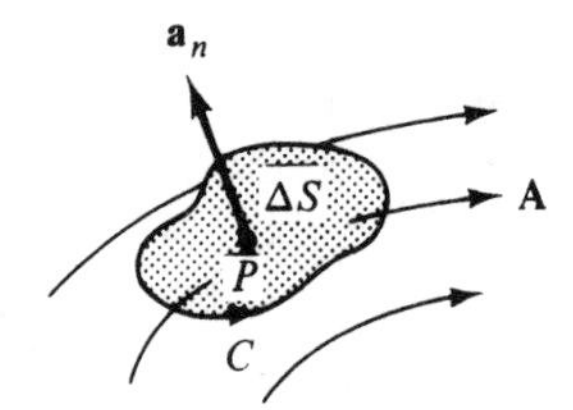

Figure 3-3 Defining the curl of **A**.

In a coordinate system, curl **A** is completely specified by its components along the coordinate unit vectors. For example, the x component in cartesian coordinates is defined by taking as the contour C a square in the x = constant plane through P as shown in Figure 3-4.

$$\nabla\times\mathbf{A}\bullet\mathbf{a}_x \equiv \lim_{\Delta y\Delta z\to 0}\frac{\oint \mathbf{A}\bullet d\mathbf{l}}{\Delta y\Delta z}$$

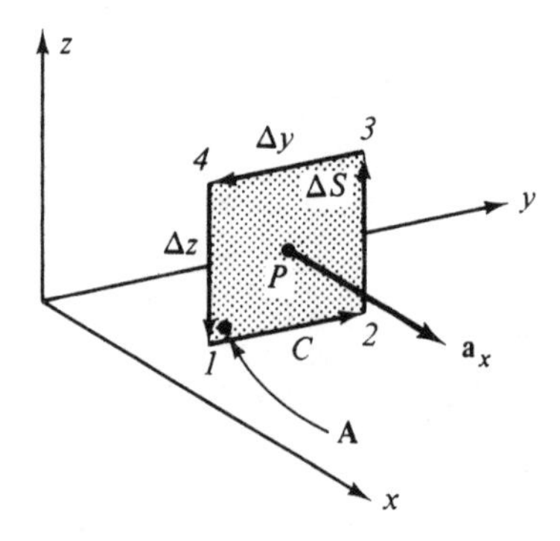

Figure 3-4 Defining the x-component of the curl **A**.

If $\mathbf{A} = A_x\mathbf{a}_x + A_y\mathbf{a}_y + A_z\mathbf{a}_z$ at the corner of ΔS closest to the origin (point *1*), then

$$\oint = \int_1^2 + \int_2^3 + \int_3^4 + \int_4^1$$

$$= A_y\Delta y + \left(A_z + \frac{\partial A_z}{\partial y}\Delta y\right)\Delta z + \left(A_y + \frac{\partial A_y}{\partial z}\Delta z\right)(-\Delta y) + A_z(-\Delta z)$$

$$= \left(\frac{\partial A_z}{\partial y} - \frac{\partial A_y}{\partial z}\right)\Delta y\,\Delta z$$

and

$$\nabla \times \mathbf{A} \bullet \mathbf{a}_x = \frac{\partial A_z}{\partial y} - \frac{\partial A_y}{\partial z}$$

The y and z components can be determined in a similar fashion. Combining the three components, the curl **A** in Cartesian coordinates is

$$\nabla \times \mathbf{A} = \left(\frac{\partial A_z}{\partial y} - \frac{\partial A_y}{\partial z}\right)\mathbf{a}_x + \left(\frac{\partial A_x}{\partial z} - \frac{\partial A_z}{\partial x}\right)\mathbf{a}_y + \left(\frac{\partial A_y}{\partial x} - \frac{\partial A_x}{\partial y}\right)\mathbf{a}_z \quad (3)$$

Expressions for the curl **A** in cylindrical and spherical coordinates can be derived in the same manner as above, though with more difficulty. In cylindrical coordinates,

$$\nabla \times \mathbf{A} = \left(\frac{1}{r}\frac{\partial A_z}{\partial \phi} - \frac{\partial A_\phi}{\partial z}\right)\mathbf{a}_r + \left(\frac{\partial A_r}{\partial z} - \frac{\partial A_z}{\partial r}\right)\mathbf{a}_\phi + \frac{1}{r}\left[\frac{\partial (rA_\phi)}{\partial r} - \frac{\partial A_r}{\partial \phi}\right]\mathbf{a}_z \quad (4)$$

In spherical coordinates,

$$\nabla \times \mathbf{A} = \frac{1}{r\sin\theta}\left[\frac{\partial (A_\phi \sin\theta)}{\partial \theta} - \frac{\partial A_\theta}{\partial \phi}\right]\mathbf{a}_r + \frac{1}{r}\left[\frac{1}{\sin\theta}\frac{\partial A_r}{\partial \phi} - \frac{\partial (rA_\phi)}{\partial r}\right]\mathbf{a}_\theta + \frac{1}{r}\left[\frac{\partial (rA_\theta)}{\partial r} - \frac{\partial A_r}{\partial \theta}\right]\mathbf{a}_\phi \quad (5)$$

Frequently useful are two properties of the curl **A**:

1. The divergence of the curl of a vector is zero.

$$\nabla \bullet (\nabla \times \mathbf{A}) = 0$$

2. The curl of the gradient of a scalar function is zero

$$\nabla \times (\nabla f) = 0$$

For example, under static conditions, the electric field $\mathbf{E} = -\nabla V$ so

$$\nabla \times \mathbf{E} = 0$$

This is another test for conservation of a vector field: if the curl is zero, the field is conservative.

In view of Ampere's law, the defining equation for (curl $\mathbf{H}$)$_x$ may be rewritten as

$$\nabla \times \mathbf{H} \bullet \mathbf{a}_x \equiv \lim_{\Delta y \Delta z \to 0} \frac{\oint \mathbf{H} \bullet d\mathbf{l}}{\Delta y \Delta z} = \lim_{\Delta y \Delta z \to 0} \frac{I_x}{\Delta y \Delta z} \equiv J_x$$

where $J_x = dI_x / dS$ is the area density of x-directed current. Thus the x components of (curl $\mathbf{H}$)$_x$ and the *current density* J_x are equal at any point. The relationship is similar for the y and z components so that the entire relation can be written as

$$\nabla \times \mathbf{H} = \mathbf{J} \tag{6}$$

This is a differential form of Ampere's law for static magnetic fields. Thus, the magnetic field $\mathbf{H}$ is not conservative.

Example 3.3 A long, straight conductor cross section with radius a has a magnetic field strength $\mathbf{H} = (Ir/2\pi a^2)\mathbf{a}_\phi$ within the conductor ($r < a$) and $\mathbf{H} = (I/2\pi r)\mathbf{a}_\phi$ for $r > a$. Find the current density $\mathbf{J}$ in both regions.

Solution: Within the conductor, using (4) and (6),

$$\mathbf{J} = \nabla \times \mathbf{H} = -\frac{\partial}{\partial z}\left(\frac{Ir}{2\pi a^2}\right)\mathbf{a}_r + \frac{1}{r}\frac{\partial}{\partial r}\left(\frac{Ir^2}{2\pi a^2}\right)\mathbf{a}_z = \frac{I}{\pi a^2}\mathbf{a}_z$$

which corresponds to a current of magnitude I in the $+z$-direction which is distributed uniformly over the cross-sectional area πa^2.

Outside the conductor,

$$\mathbf{J} = \nabla \times \mathbf{H} = -\frac{\partial}{\partial z}\left(\frac{I}{2\pi r}\right)\mathbf{a}_r + \frac{1}{r}\frac{\partial}{\partial r}\left(\frac{I}{2\pi}\right)\mathbf{a}_z = 0$$

which means the current is only flowing in the conductor which makes physical sense.

Magnetic Flux Density and Gauss' Law

Like **D**, the magnetic field strength **H** depends only on (moving) charges and is independent of medium. The force field associated with **H** is the *magnetic flux density* **B**, which is given by

$$\mathbf{B} = \mu \mathbf{H} \tag{7}$$

where $\mu = \mu_0 \mu_r$ is the *permeability* of the medium. The unit of **B** is the *tesla*,

$$1\,\text{T} = 1\frac{\text{N}}{\text{A} \bullet \text{m}}$$

The free space permeability μ_0 has a numerical value of $4\pi \times 10^{-7}$ and has unit *henry per meter*, H/m. *Non-magnetic* materials have *relative permeability* μ_r close to unity while *magnetic* materials (e.g., iron, ferromagnetics) can have μ_r much greater than unity.

Magnetic flux, Φ, through a surface is defined as

$$\Phi = \int_S \mathbf{B} \bullet d\mathbf{S} \tag{8}$$

The sign on Φ may be positive or negative depending upon the choice of surface normal in $d\mathbf{S}$. The unit of magnetic flux is the *weber*, Wb. The various magnetic units are related by

$$1\,\text{T} = 1\,\text{Wb/m}^2, \qquad 1\,\text{H} = 1\,\text{Wb/A}$$

Example 3.4 Find the flux crossing the portion of the plane $\phi = \pi/4$ defined by $0.01 < r < 0.05$ m and $0 < z < 2$ m (see Figure 3-5). A current filament of 2.50 A along the z-axis is in the $\mathbf{a}_z$-direction.

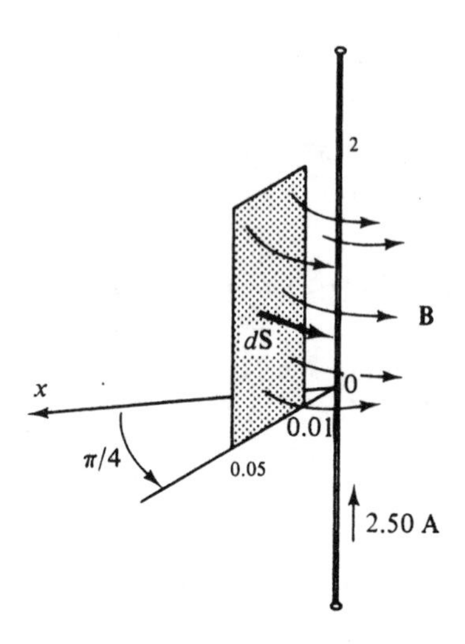

Figure 3-5 Magnetic flux from a current through a rectangle.

Solution: Using Figure 3-5, the results of Example 3.2 and (7) give the magnetic flux density

$$\mathbf{B} = \mu_0 \mathbf{H} = \frac{\mu_0 I}{2\pi r} \mathbf{a}_\phi$$

From the figure,

$$d\mathbf{S} = d_r d_z \mathbf{a}_\phi$$

The magnetic flux passing through the rectangle is then [using (8)]

$$\Phi = \int_0^2 \int_{0.01}^{0.05} \frac{\mu_0 I}{2\pi r} \mathbf{a}_\phi \bullet dr\, dz\, \mathbf{a}_\phi$$

$$= \frac{2\mu_0 I}{2\pi} \ln \frac{0.05}{0.01} = 1.61 \times 10^{-6} \text{ Wb} = 1.61\ \mu\text{Wb}$$

It should be observed that the lines of magnetic flux Φ are closed curves, with no starting point or termination. They are said to be *solenoidal.* This is in contrast with electric flux, Ψ, which originates on positive charge and terminates on negative charge. In Figure 3-6, all of the magnetic flux that enters the closed surface must leave the closed surface. Thus **B** fields have no sources or sinks, which is mathematically expressed by

$$\nabla \bullet \mathbf{B} = 0 \tag{9}$$

NOTE! Equation (9) is referred to as *Gauss' law for magnetic fields.*

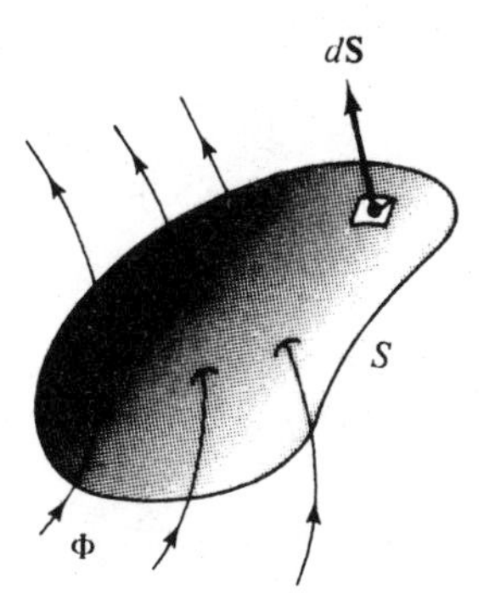

Figure 3-6 Closed surface in the presence of **B**.

Inductance

The *inductance L* of a conductor system may be defined as *the ratio of the linking magnetic flux to the current producing the flux.* For static (or, at most, low frequency) current I and a coil containing N turns, as shown in Figure 3-7,

$$L = \frac{N\Phi}{I} \tag{10}$$

where Φ is the flux passing through a single loop. The unit on L is *henry*, where 1 H = 1 Wb/A.

Inductance is also given by

$$L = \frac{\lambda}{I} \tag{11}$$

where λ, *the flux linkage*, $N\Phi$ for coils with N turns or simply Φ for other conductor arrangements.

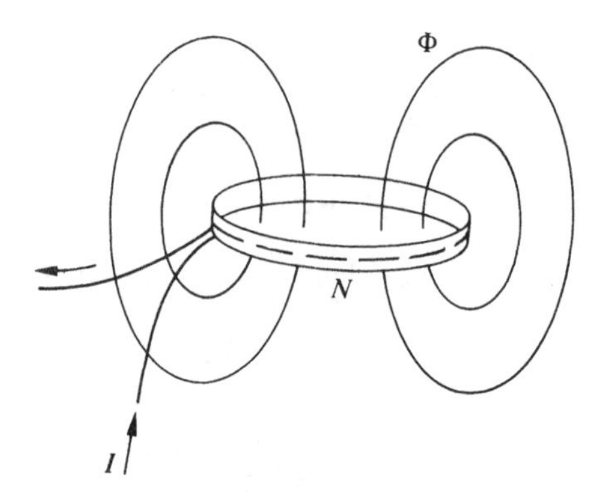

Figure 3-7 Flux linkage for a current-carrying coil.

It should also be noted that L will always be the product of the permeability of the medium μ and a geometrical factor having units of length.

Example 3.5 Find the inductance per unit length of a coaxial conductor such as that shown in Figure 3-8.

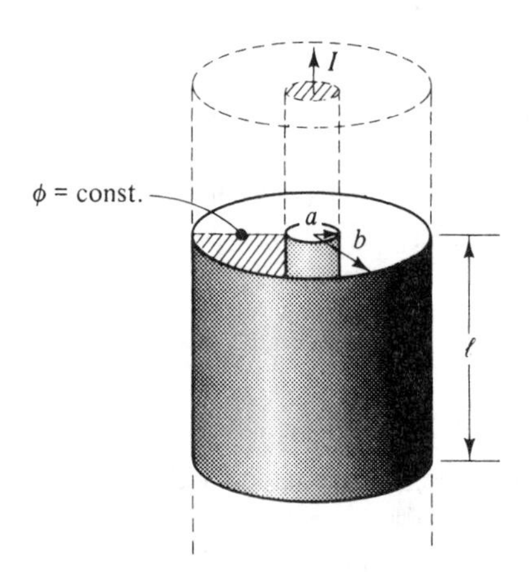

Figure 3-8 Coaxial conductor.

Solution: Between the conductors, the magnetic fields are given by

$$\mathbf{H} = \frac{I}{2\pi r}\mathbf{a}_{\phi}$$

$$\mathbf{B} = \frac{\mu_0 I}{2\pi r} \mathbf{a}_\phi$$

The currents in the two conductors are linked by the flux across the surface ϕ = const. For a length of $\ell = 1$ m,

$$\lambda = \int_0^1 \int_a^b \frac{\mu_0 I}{2\pi r} dr\, dz = \frac{\mu_0 I}{2\pi} \ln \frac{b}{a}$$

Using (11), the per unit length inductance of the coaxial conductor is

$$L \text{ per meter} = \frac{\mu_0}{2\pi} \ln \frac{b}{a} \quad (\text{H / m})$$

Figures 3-9 through 3-13 give exact and/or approximate inductances of some common noncoaxial conductor arrangements.

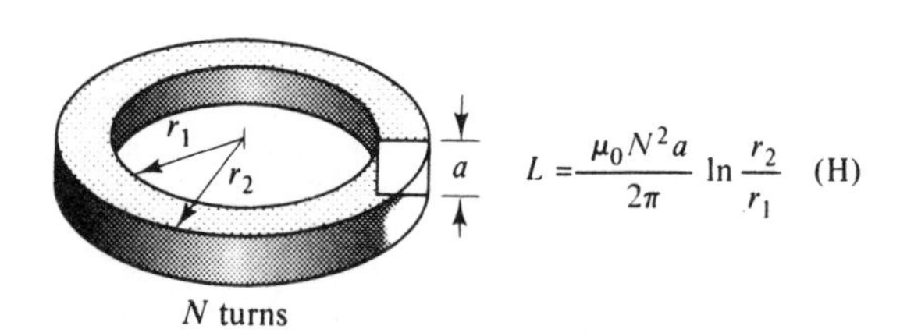

Figure 3-9 Toroid, square cross section.

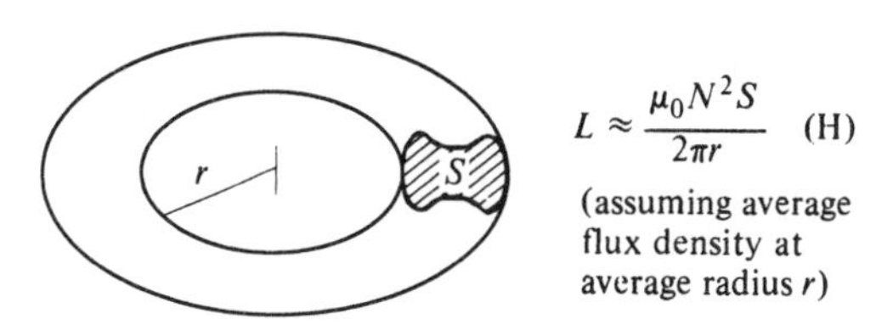

Figure 3-10 Toroid, general cross section S.

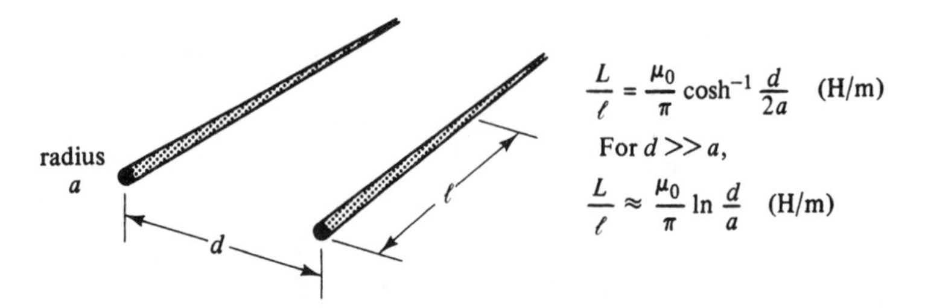

Figure 3-11 Parallel conductors of radius *a*.

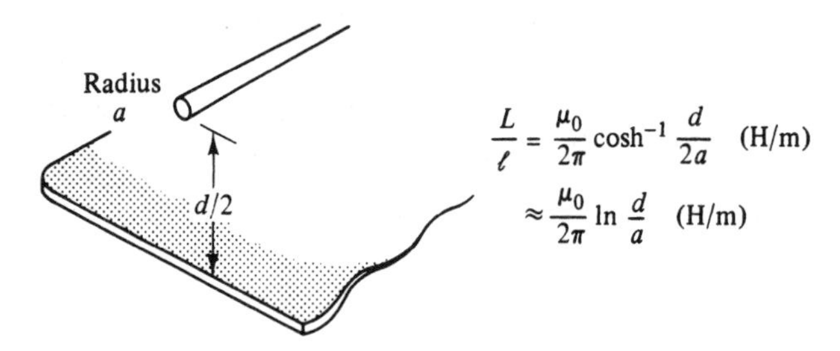

Figure 3-12 Cylindrical conductor parallel to a ground plane.

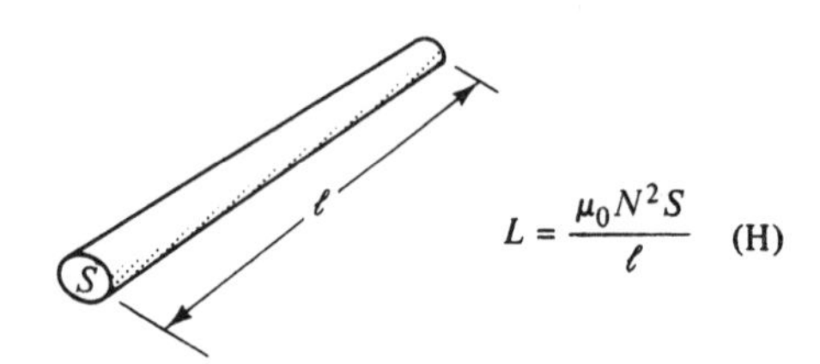

Figure 3-13 Long solenoid of small cross-sectional area *S*.

Important Things to Remember:

- ✔ The magnetic fields **H** and **B** circulate around a conducting wire carrying current I according to a right-hand rule.
- ✔ In an isotropic medium, $\mathbf{B} = \mu\mathbf{H}$.
- ✔ Magnetic flux lines are solenoidal meaning that they are closed curves with no beginning or end.
- ✔ For a closed surface, the total magnetic flux entering the surface is equal to the total magnetic flux leaving the surface.
- ✔ The inductance of a conductor is the magnetic flux linkage per unit current.

Solved Problems

Solved Problem 3.1

Find **H** in the center of a square loop of side *L*.

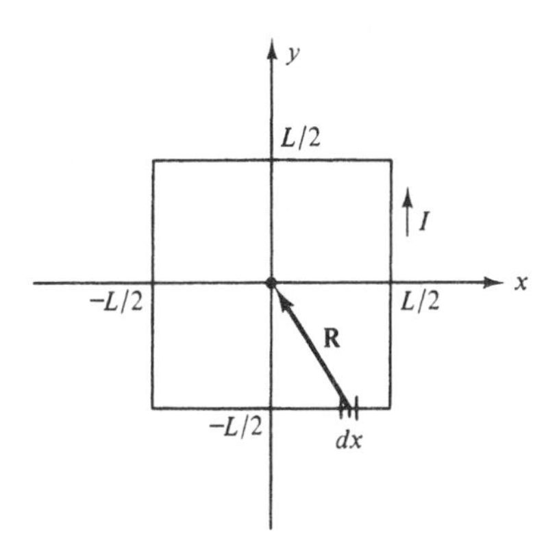

Figure 3-14 Square loop of current *I*.

Solution: Choose a Cartesian coordinate system such that the loop is centered as shown in Figure 3-14. By symmetry, each half-side contributes the same amount of **H** at the center. For the half-side, $0 \le x \le L/2$, $y = -L/2$, and the Biot-Savart law gives for the field at the origin

$$d\mathbf{H} = \frac{(I\,dx\,\mathbf{a}_x)\times[-x\mathbf{a}_x + (L/2)\mathbf{a}_y]}{4\pi[x^2 + (L/2)^2]^{3/2}}$$

$$= \frac{I\,dx(L/2)\mathbf{a}_z}{4\pi[x^2 + (L/2)^2]^{3/2}} \text{ A/m}$$

Therefore, the total field at the origin is

$$\mathbf{H} = 8\int_0^{L/2} \frac{I\,dx(L/2)\mathbf{a}_z}{4\pi[x^2 + (L/2)^2]^{3/2}}$$

$$= \frac{2\sqrt{2}I}{\pi L}\mathbf{a}_z = \frac{2\sqrt{2}I}{\pi L}\mathbf{a}_n \text{ A/m}$$

where $\mathbf{a}_n$ is the unit normal to the plane of the loop as given by the usual right-hand rule.

Solved Problem 3.2

A thin cylindrical conductor of radius a, infinite in length, carries a current I. Find **H** at all points using Ampere's law.

Solution: The Biot-Savart law shows that **H** has only a ϕ-component. Furthermore, H_ϕ is a function of r only. Proper paths for Ampere's law are concentric circles. For path *1* shown in Figure 3-15,

$$\oint \mathbf{H}\bullet d\mathbf{l} = 2\pi\,r\,H_\phi = I_{enc} = 0$$

and for path *2*,

$$\oint \mathbf{H}\bullet d\mathbf{l} = 2\pi\,r\,H_\phi = I$$

Thus, for points within the cylindrical conducting shell, $\mathbf{H} = 0$, and for external points, $\mathbf{H} = (I/2\pi\,r)\,\mathbf{a}_\phi$ A/m. For $r > a$, this is the same field as that of a current filament I along the axis.

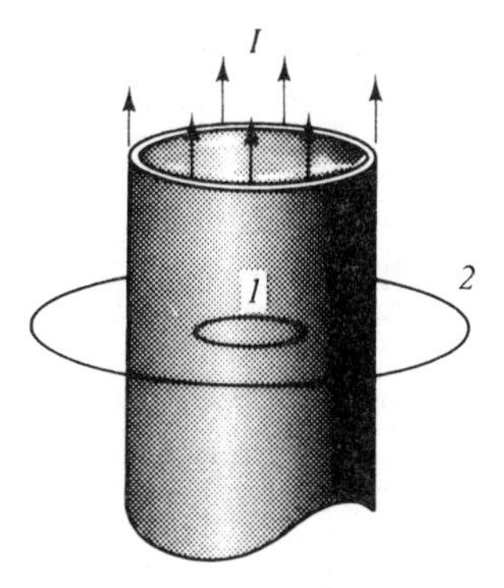

Figure 3-15 Current-carrying cylindrical shell.

Solved Problem 3.3

A radial field

$$\mathbf{H} = \frac{2.39 \times 10^6}{r} \cos\phi \, \mathbf{a}_r \ \text{A/m}$$

exists in free space. Find the magnetic flux Φ crossing the surface defined by $-\pi/4 \leq \phi \leq \pi/4$, $0 \leq z \leq 1$ m. See Figure 3-16.

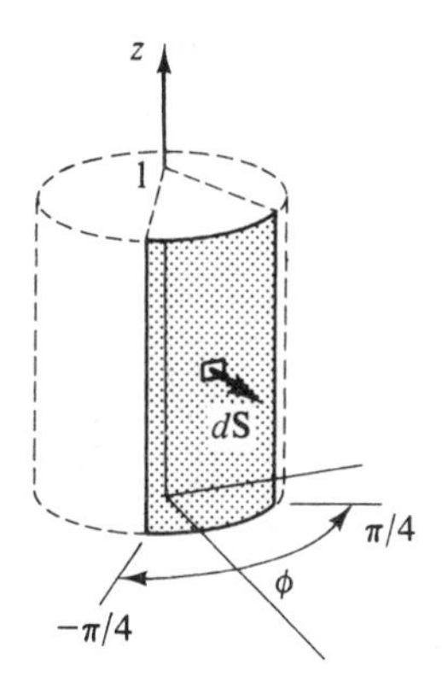

Figure 3-16 Magnetic flux passing through cylindrical surface.

Solution: The flux density in free space is

$$\mathbf{B} = \mu_0 \mathbf{H} = \frac{3.00}{r} \cos\phi \, \mathbf{a}_r \ \ \text{T}$$

and the flux passing through the surface is

$$\Phi = \int_0^1 \int_{-\pi/4}^{\pi/4} \left(\frac{3.00}{r} \cos\phi \, \mathbf{a}_r \right) \bullet (r \, d\phi \, dz \, \mathbf{a}_r) = 4.24 \ \text{Wb}$$

Solved Problem 3.4

Find the inductance per unit length of the parallel cylindrical conductors shown in Figure 3-11, where $d = 25$ feet and $a = 0.803$ inch.

Solution: Using the formulas from Figure 3-11,

$$\frac{L}{\ell} = \frac{\mu_0}{\pi} \cosh^{-1} \frac{d}{2a} = (4 \times 10^{-7}) \cosh^{-1} \frac{25(12)}{2(0.803)} = 2.37 \ \mu\text{H/m}$$

The approximate formula gives

$$\frac{L}{\ell} \approx \frac{\mu_0}{\pi} \ln \frac{d}{a} = 2.37 \ \mu\text{H/m}$$

When $d/a \geq 10$, the approximate formula may be used with an error of less than 0.5 %.

Solved Problem 3.5

Assume that the air-core toroid shown in Figure 3-9 has 700 turn, an inner radius of 1 cm, an outer radius of 2 cm, and height $a = 1.5$ cm. Find L using (*a*) the formula for square cross-section toroids; (*b*) the approximate formula for a general toroid, which assumes a uniform **H** at mean radius (Figure 3-10).

Solution: (*a*) For the square cross-section, from Figure 3-9,

$$L = \frac{\mu_0 N^2 a}{2\pi} \ln \frac{r_2}{r_1} = \frac{(4\pi \times 10^{-7})(700)^2(0.015)}{2\pi} \ln 2 = 1.02 \ \text{mH}$$

(*b*) Using the approximate formula from Figure 3-10,

$$L = \frac{\mu_0 N^2 S}{2\pi r} = \frac{(4\pi \times 10^{-7})(700)^2 (0.01)(0.015)}{2\pi(0.015)} = 0.98 \text{ mH}$$

With a radius r that is larger compared to the cross section, the two formulas yield even closer results.

Chapter 4
TIME-VARYING FIELDS AND MAXWELL'S EQUATIONS

IN THIS CHAPTER:

- ✔ *Faraday's Law and Induced EMF*
- ✔ *Ampere's Law and Displacement Current*
- ✔ *Boundary Conditions*
- ✔ *Maxwell's Equations*
- ✔ *Solved Problems*

Chapters 2 and 3 dealt with fields that do not vary with time. For time-varying fields, the equations for **E** and **H** need to be modified. Collectively the set of equations relating the time-varying **E** and **H** are called *Maxwell's equations*.

Faraday's Law and Induced EMF

An electric current is one source for a magnetic field. Faraday observed that a time-varying magnetic field can induce a current in a closed loop of wire. This induced current is driven by an *electromotive force* (*emf*), which is effectively a voltage around the loop.

The relationship between the emf and the time-varying magnetic field is governed by the integral form of *Faraday's law*:

$$emf = \oint_C \mathbf{E} \bullet d\mathbf{l} = -\frac{d}{dt}\int_S \mathbf{B} \bullet d\mathbf{S} \tag{1}$$

The positive sense around the loop contour C and the direction of the surface normal $d\mathbf{S}$ are related by another right-hand rule: If the fingers of the right hand curl in the direction of integration along C, the thumb of the right hand points in the direction of the normal vector [Figure 4-1(*a*)]. Now if **B** is increasing with time, the time derivative will be positive and, thus, the right side of (1) will be negative. In order for the left integral to be negative, the direction of **E** must be opposite to that of the contour [Figure 4-1(*b*)]. A conducting filament in place of the contour would carry a current i_c also in the direction of **E**. As shown in Figure 4-1(*c*), such a current loop generates an induced magnetic flux ϕ' which opposes the increase in **B**. This is summarized by *Lenz's law*:

Note!

The voltage induced by a changing flux has a polarity such that the induced current in the closed path gives rise to a secondary magnetic flux which opposes the change in time-varying source magnetic flux.

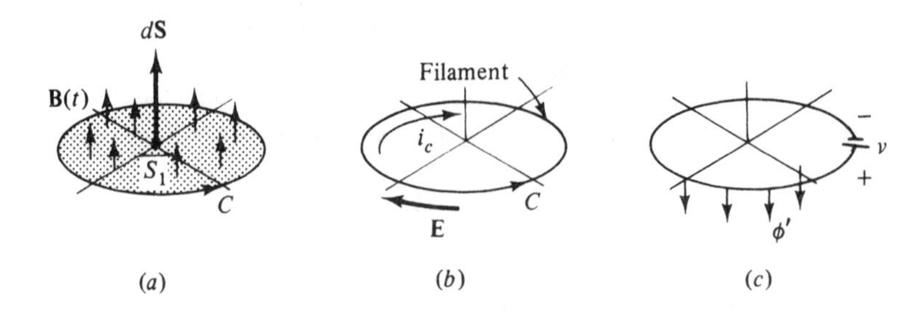

Figure 4-1 Illustration of Lenz's law.

Example 4.1 The circular loop conductor in Figure 4-2 lies in the $z = 0$ plane. The loop has a radius of 0.10 m and a resistance R of 5 Ω. Given $\mathbf{B} = 0.20 \sin 10^3 t\, \mathbf{a}_z$ (T), determine the current.

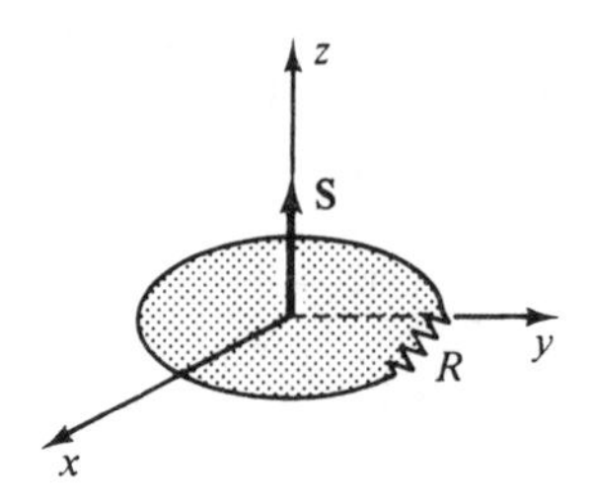

Figure 4-2 Conducting loop with a resistor

Solution: The total flux does not depend on x or y so the total flux passing through the loop is

$$\Phi = \mathbf{B} \bullet \mathbf{S} = 2 \times 10^{-3} \pi \sin 10^3 t \quad \text{(Wb)}$$

The voltage (emf) around the loop is

$$v = -\frac{d\Phi}{dt} = -2\pi \cos 10^3 t \quad \text{(V)}$$

and the current is

$$i = \frac{v}{R} = -0.4\pi \cos 10^3 t \quad \text{(A)}$$

Ampere's Law and Displacement Current

In static fields, the curl of **H** was found to be point-wise equal to the current density $\mathbf{J}_c$, the *conduction current density*; the subscript c was added to emphasize that charge in motion composes that current. If the divergence is applied to the curl, a vector identity requires that

$$\nabla \bullet \nabla \times \mathbf{H} = 0 \tag{2}$$

However, the divergence of $\mathbf{J}_c$ is nonzero for time-varying fields. This relation is described by the *continuity equation*

$$\nabla \bullet \mathbf{J}_c = -\frac{\partial \rho}{\partial t} \tag{3}$$

The continuity equation is a statement of charge conservation. In words, the rate of movement of charge out of a region is equal to the time rate of decrease of charge contained within the region. Hence, Maxwell postulated that

$$\nabla \times \mathbf{H} = \mathbf{J}_c + \mathbf{J}_D \tag{4}$$

where $\mathbf{J}_D$ is the *displacement current density*. Taking the divergence of (4) and using (2) and (3) gives

$$\nabla \bullet \mathbf{J}_c = -\nabla \bullet \mathbf{J}_D = -\frac{\partial \rho}{\partial t}$$

Using Gauss' law [(10) from Chapter 2],

$$-\nabla \bullet \mathbf{J}_D = -\frac{\partial \nabla \bullet \mathbf{D}}{\partial t} \quad \Rightarrow \quad \mathbf{J}_D = \frac{\partial \mathbf{D}}{\partial t} \tag{5}$$

Using (5) in (4) gives a point form for Ampere's law

$$\nabla \times \mathbf{H} = \mathbf{J}_c + \frac{\partial \mathbf{D}}{\partial t} \tag{6}$$

An integral form of Ampere's law that is valid for time-varying fields is obtained using *Stoke's theorem*. Stoke's theorem states that

$$\oint_C \mathbf{F} \bullet d\mathbf{l} = \int_S \nabla \times \mathbf{F} \bullet d\mathbf{S}$$

where the open surface S is bounded by the contour C. Integrating (6) over a fixed open surface gives

$$\int_S \nabla \times \mathbf{H} \bullet d\mathbf{S} = \oint_C \mathbf{H} \bullet d\mathbf{l}$$

$$\Rightarrow \oint_C \mathbf{H} \bullet d\mathbf{l} = \int_S \mathbf{J}_c \bullet d\mathbf{S} + \frac{d}{dt}\int_S \mathbf{D} \bullet d\mathbf{S} \qquad (7)$$

The first term on the right-hand side of (7) is the conduction current i_c. The last term in (7) is the displacement current i_D through a fixed surface S is obtained by integrating the normal component of $\mathbf{J}_D$ over the surface (just as i_c is obtained from $\mathbf{J}_c$)

$$i_D = \int_S \mathbf{J}_D \bullet d\mathbf{S} = \int_S \frac{\partial \mathbf{D}}{\partial t} \bullet d\mathbf{S} = \frac{d}{dt}\int_S \mathbf{D} \bullet d\mathbf{S} \qquad (8)$$

Example 4.2 Show that $i_c = i_D$ in the circuit of Figure 4-3.

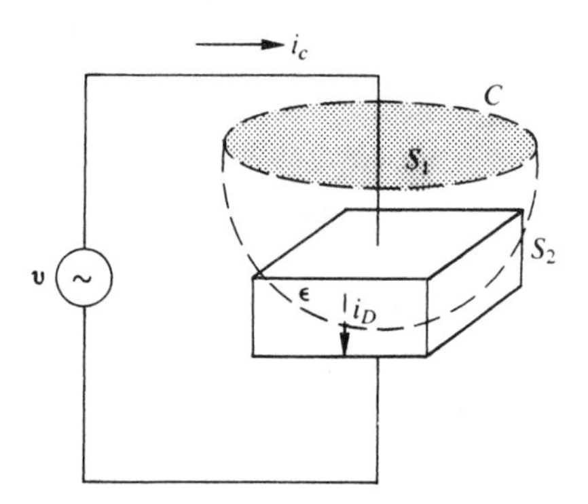

Figure 4-3 Time-varying capacitor circuit.

Solution: Since the two surfaces S_1 and S_2 have the common boundary contour C,

$$\oint_C \mathbf{H} \bullet d\mathbf{l} = \int_{S_1} \mathbf{J}_c \bullet d\mathbf{S} + \frac{d}{dt}\int_{S_1} \mathbf{D} \bullet d\mathbf{S} = \int_{S_2} \mathbf{J}_c \bullet d\mathbf{S} + \frac{d}{dt}\int_{S_2} \mathbf{D} \bullet d\mathbf{S}$$

Assuming the flux of the capacitor is confined to the dielectric between the conducting plates, $\mathbf{D} = 0$ over S_1. And, since there are no free charges present in the dielectric, $\mathbf{J}_c = 0$ over S_2. Therefore,

$$\int_{S_1} \mathbf{J}_c \bullet d\mathbf{S} = \frac{d}{dt}\int_{S_2} \mathbf{D} \bullet d\mathbf{S} = \int_{S_2} \frac{\partial \mathbf{D}}{\partial t} \bullet d\mathbf{S} \quad \text{or} \quad i_c = i_D$$

It should be noted that $\partial \mathbf{D} / \partial t$ is nonzero only over that part of S_2 that lies within the dielectric.

Example 4.3 Repeat Example 4.2, this time using circuit analysis.

Solution: The capacitance of the capacitor is

$$C = \frac{\varepsilon A}{d}$$

where A is the area of the plates and d is the separation. The conduction current is then

$$i_c = C\frac{\partial v}{\partial t} = \frac{\varepsilon A}{d}\frac{\partial v}{\partial t}$$

On the other hand, the electric field in the dielectric is, neglecting fringing, $E = v/d$. Hence,

$$D = \varepsilon E = \frac{\varepsilon}{d}v, \qquad \frac{\partial \mathbf{D}}{\partial t} = \frac{\varepsilon}{d}\frac{\partial v}{\partial t}$$

and the displacement is [using (8); $\mathbf{D}$ is normal to the plates]

$$i_D = \int_A \frac{\partial \mathbf{D}}{\partial t} \bullet d\mathbf{S} = \int_A \frac{\varepsilon}{d}\frac{\partial v}{\partial t} dS = \frac{\varepsilon A}{d}\frac{\partial v}{\partial t} = i_c$$

Boundary Conditions

If the conductor in Figures 2-23 and 2-24 is replaced with a second, different dielectric, then the same argument as was made for the conductor-dielectric boundary in Chapter 2 gives the following two electric field boundary conditions:

1. The *tangential component of* **E** *is continuous across a dielectric interface*. In symbols,

$$E_{t1} = E_{t2} \qquad \text{and} \qquad \frac{D_{t1}}{\varepsilon_{r1}} = \frac{D_{t2}}{\varepsilon_{r2}} \tag{9}$$

2. *The normal component of* **D** *has a discontinuity of magnitude* $|\rho_s|$ *across a dielectric interface*. If the unit normal is chosen to point into dielectric 2, then this condition can be written

$$D_{n2} - D_{n1} = \rho_s \qquad \text{and} \qquad \varepsilon_{r2}E_{n2} - \varepsilon_{r1}E_{n1} = \frac{\rho_s}{\varepsilon_0} \tag{10}$$

Generally the interface will have no free charges so that

$$D_{n2} = D_{n1} \qquad \text{and} \qquad \varepsilon_{r2}E_{n2} = \varepsilon_{r1}E_{n1} \tag{11}$$

Example 4.4 Given that $\mathbf{E}_1 = 2\mathbf{a}_x - 3\mathbf{a}_y + 5\mathbf{a}_z$ V/m at the charge-free interface of Figure 4-4, find $\mathbf{D}_2$ and the angles θ_1 and θ_2.

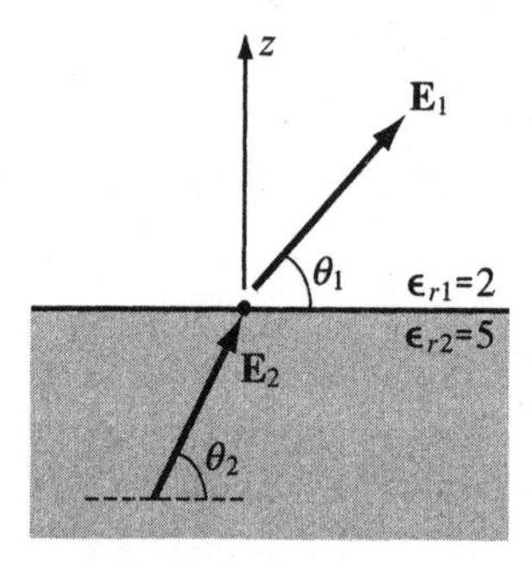

Figure 4-4 Dielectric-dielectric interface.

Solution: The interface is a z = constant plane. The x and y components are tangential and the z components are normal. By continuity of tangential components of **E** and normal component of **D**,

$$\begin{aligned}
\mathbf{E}_1 &= 2\mathbf{a}_x - 3\mathbf{a}_y + 5\mathbf{a}_z \\
\mathbf{E}_2 &= 2\mathbf{a}_x - 3\mathbf{a}_y + E_{2z}\mathbf{a}_z \\
\mathbf{D}_1 &= \varepsilon_0\varepsilon_{r1}\mathbf{E}_1 = 4\varepsilon_0\mathbf{a}_x - 6\varepsilon_0\mathbf{a}_y + 10\varepsilon_0\mathbf{a}_z \\
\mathbf{D}_2 &= \varepsilon_0\varepsilon_{r2}\mathbf{E}_2 = D_{2x}\mathbf{a}_x + D_{2y}\mathbf{a}_y + 10\varepsilon_0\mathbf{a}_z
\end{aligned}$$

The unknowns are now found from the relation $\mathbf{D}_2 = \varepsilon_0 \varepsilon_{r2} \mathbf{E}_2$

$$D_{2x}\mathbf{a}_x + D_{2y}\mathbf{a}_y + 10\varepsilon_0\mathbf{a}_z = 2\varepsilon_0\varepsilon_{r2}\mathbf{a}_x - 3\varepsilon_0\varepsilon_{r2}\mathbf{a}_y + \varepsilon_0\varepsilon_{r2}E_{2z}\mathbf{a}_z$$

from which

$$D_{2x} = 2\varepsilon_0\varepsilon_{r2} = 10\varepsilon_0, \quad D_{2y} = -3\varepsilon_0\varepsilon_{r2} = -15\varepsilon_0, \quad E_{2z} = \frac{10}{\varepsilon_{r2}} = 2$$

The angles made with the plane of the interface are most easily found from

$$\mathbf{E}_1 \bullet \mathbf{a}_z = |\mathbf{E}_1|\cos(90° - \theta_1) \Rightarrow 5 = \sqrt{38}\sin\theta_1 \Rightarrow \theta_1 = 54.2°$$
$$\mathbf{E}_2 \bullet \mathbf{a}_z = |\mathbf{E}_2|\cos(90° - \theta_2) \Rightarrow 2 = \sqrt{17}\sin\theta_2 \Rightarrow \theta_2 = 29.0°$$

For magnetic fields, consider the interface of Figure 4-5 showing the boundary between material 1 and material 2.

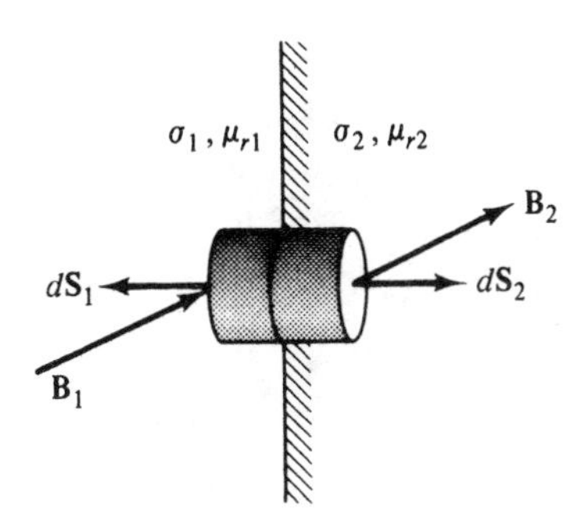

Figure 4-5 Closed surface at the boundary between two materials.

The behavior of normal **B** can be determined by use of a small right circular cylinder positioned across the interface as shown. Since magnetic flux lines are continuous,

$$\oint \mathbf{B} \bullet d\mathbf{S} = \int_{end1} \mathbf{B}_1 \bullet d\mathbf{S}_1 + \int_{curved\,side} \mathbf{B} \bullet d\mathbf{S} + \int_{end2} \mathbf{B}_2 \bullet d\mathbf{S}_2 = 0$$

Now, if the two ends are allowed to approach one another, keeping the interface between the ends, the area of the curved surface goes to zero giving

$$\int_{end\,1} \mathbf{B}_1 \bullet d\mathbf{S}_1 + \int_{end\,2} \mathbf{B}_2 \bullet d\mathbf{S}_2 = 0$$

$$-B_{n1} \int_{end\,1} dS_1 + B_{n2} \int_{end\,2} dS_2 = 0$$

$$\Rightarrow B_{n1} = B_{n2} \qquad (12)$$

In words, *the normal component of* **B** *is continuous across the interface.*

The variation in tangential **H** across an interface is obtained by the application of Ampere's law around the closed rectangular path shown in Figure 4-6.

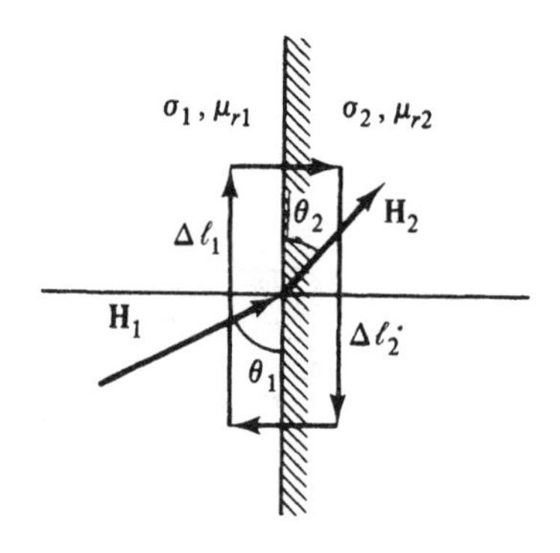

Figure 4-6 Closed path at the boundary between two materials.

Assuming no current at the interface and letting the rectangle shrink down on to the boundary,

$$\oint \mathbf{H} \bullet d\mathbf{l} \rightarrow H_{t1}\Delta\ell_1 - H_{t2}\Delta\ell_2 = H_{t1}\Delta\ell - H_{t2}\Delta\ell$$

where $\Delta\ell_1 = \Delta\ell_2 = \Delta\ell$

For the displacement current term, S is the area bounded by the rectangular contour. When shrinking the rectangle, the area goes to zero giving

$$\frac{d}{dt}\int_S \mathbf{D} \bullet d\mathbf{S} \to 0$$

For the conduction current, there are configurations that support a current sheet **K** (A/m) at the boundary (Figure 4-7). In that regard, the shrinking area of the rectangle always contains the surface current. The conduction current term then gives

$$\int_S \mathbf{K} \bullet d\mathbf{S} \to K\Delta\ell$$

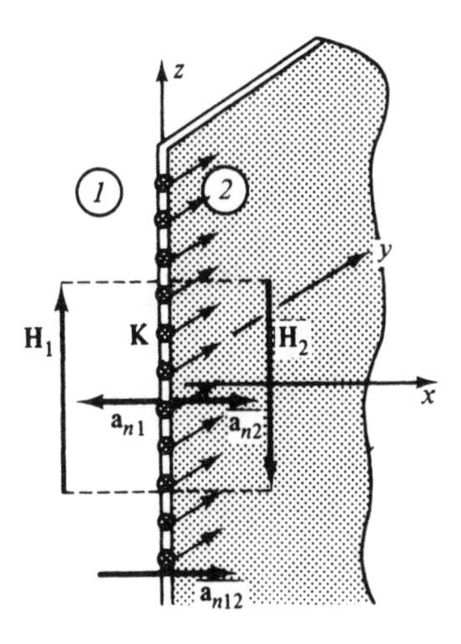

Figure 4-7 Boundary with a surface current.

Equating both sides of Ampere's law using the above gives

$$H_{t1}\Delta\ell - H_{t2}\Delta\ell = K\Delta\ell \text{ or } H_{t1} - H_{t2} = K$$

Since **K** is a vector, this is better expressed using a vector expression giving the directions

$$(\mathbf{H}_1 - \mathbf{H}_2) \times \mathbf{a}_{n12} = \mathbf{K} \qquad (13)$$

where the normal unit vector $\mathbf{a}_{n12}$ is directed from region 1 into region 2. For the case of a non-conducting dielectric-dielectric interface (which does not support a surface current), $\mathbf{K} = 0$ and the tangential **H** is continuous across the boundary

$$H_{t1} = H_{t2} \tag{14}$$

Example 4.5 Region 1 is defined for $x < 0$ and has relative permeability of $\mu_{r1} = 3$. Region 2 is defined for $x > 0$ and has relative permeability of $\mu_{r2} = 5$. Neither region is conducting. Given

$$\mathbf{H}_1 = 4.0\mathbf{a}_x + 3.0\mathbf{a}_y - 6.0\mathbf{a}_z \text{ (A / m)}$$

find $\mathbf{H}_2$ and $\mathbf{B}_2$.

Solution: Since the boundary is $x = 0$, x components are normal and the y and z components are tangential. Using $\mathbf{B} = \mu\mathbf{H}$,

$$\mathbf{B}_1 = 12.0\mu_0\mathbf{a}_x + 9.0\mu_0\mathbf{a}_y - 18.0\mu_0\mathbf{a}_z \text{ (T)}$$

For non-conducting materials, the tangential $\mathbf{H}$ and normal $\mathbf{B}$ are continuous

$$\mathbf{H}_2 = H_{2x}\mathbf{a}_x + 3.0\mathbf{a}_y - 6.0\mathbf{a}_z \text{ (A / m)}$$
$$\mathbf{B}_2 = 12.0\mu_0\mathbf{a}_x + B_{2y}\mathbf{a}_y - B_{2z}\mathbf{a}_z \text{ (T)}$$

The unknown terms are:

$$H_{2x} = \frac{B_{2x}}{\mu_{r2}\mu_0} = 2.4, \quad B_{2y} = \mu_{r2}\mu_0 H_{2y} = 15.0$$
$$B_{2z} = \mu_{r2}\mu_0 H_{2z} = -30\mu_0$$

Maxwell's Equations

Collectively, Faraday's law, Ampere's law (with displacement current), and Gauss' laws for electric and magnetic fields are known as *Maxwell's equations*. In Table 4-1, the most general form is presented, where charges and conduction current may be present in the region. For free space and other non-conducting materials (conductivity $\sigma = 0$), where there are no charges ($\rho = 0$) and no conduction currents ($\mathbf{J}_c = 0$), Maxwell's equations take the form shown in Table 4-2.

Table 4-1. Maxwell's Equations, General Set

Point Form	Integral Form
$\nabla \times \mathbf{H} = \mathbf{J}_c + \dfrac{\partial \mathbf{D}}{\partial t}$	$\oint \mathbf{H} \cdot d\mathbf{l} = \int_S \left(\mathbf{J}_c + \dfrac{\partial \mathbf{D}}{\partial t} \right) \cdot d\mathbf{S}$ (Ampere's law)
$\nabla \times \mathbf{E} = -\dfrac{\partial \mathbf{B}}{\partial t}$	$\oint \mathbf{E} \cdot d\mathbf{l} = \int_S \left(-\dfrac{\partial \mathbf{B}}{\partial t} \right) \cdot d\mathbf{S}$ (Faraday's law; S fixed)
$\nabla \cdot \mathbf{D} = \rho$	$\oint_S \mathbf{D} \cdot d\mathbf{S} = \int_v \rho \, dv$ (Gauss' law)
$\nabla \cdot \mathbf{B} = 0$	$\oint_S \mathbf{B} \cdot d\mathbf{S} = 0$ (nonexistence of monopole)

Table 4-2. Maxwell's Equations, Free-Space Set

Point Form	Integral Form
$\nabla \times \mathbf{H} = \dfrac{\partial \mathbf{D}}{\partial t}$	$\oint \mathbf{H} \cdot d\mathbf{l} = \int_S \left(\dfrac{\partial \mathbf{D}}{\partial t} \right) \cdot d\mathbf{S}$
$\nabla \times \mathbf{E} = -\dfrac{\partial \mathbf{B}}{\partial t}$	$\oint \mathbf{E} \cdot d\mathbf{l} = \int_S \left(-\dfrac{\partial \mathbf{B}}{\partial t} \right) \cdot d\mathbf{S}$
$\nabla \cdot \mathbf{D} = 0$	$\oint_S \mathbf{D} \cdot d\mathbf{S} = 0$
$\nabla \cdot \mathbf{B} = 0$	$\oint_S \mathbf{B} \cdot d\mathbf{S} = 0$

Important Things to Remember:

- ✔ A time-varying **B**-field can induce a current in a conducting loop.
- ✔ In Ampere's law, Maxwell added the displacement current that satisfied charge conservation.
- ✔ The tangential **E** is always continuous at the boundary between two materials.
- ✔ The normal **B** is always continuous at the boundary between two materials.

Solved Problems

Solved Problem 4.1 In a material for which $\sigma = 5.0$ S/m and $\varepsilon_r = 1$, the electric field intensity is $E = 250 \sin 10^{10} t$ V/m. Find the conduction current and displacement current densities and the frequency at which they will have equal magnitudes.

Solution: The conduction current density is

$$J_c = \sigma E = 1250 \sin 10^{10} t \quad \mathrm{A/m^2}$$

On the assumption that the field direction does not vary with time,

$$J_D = \frac{\partial \mathbf{D}}{\partial t} = \frac{\partial}{\partial t}(\varepsilon_0 250 \sin 10^{10} t) = 22.1 \cos 10^{10} t \quad \mathrm{A/m^2}$$

For $J_c = J_D$, we want $\sigma = \omega\varepsilon$ or

$$\omega = \frac{5.0}{8.854 \times 10^{-12}} = 5.65 \times 10^{11} \text{ rad/s} = 89.9 \text{ GHz}$$

Solved Problem 4.2 An area of 0.65 m^2 in the $z = 0$ plane is enclosed by a filamentary conductor. Find the induced voltage given that

$$\mathbf{B} = 0.05 \cos 10^3 t \left(\frac{\mathbf{a}_y + \mathbf{a}_z}{\sqrt{2}} \right) \quad \text{T}$$

Solution: See Figure 4-8.

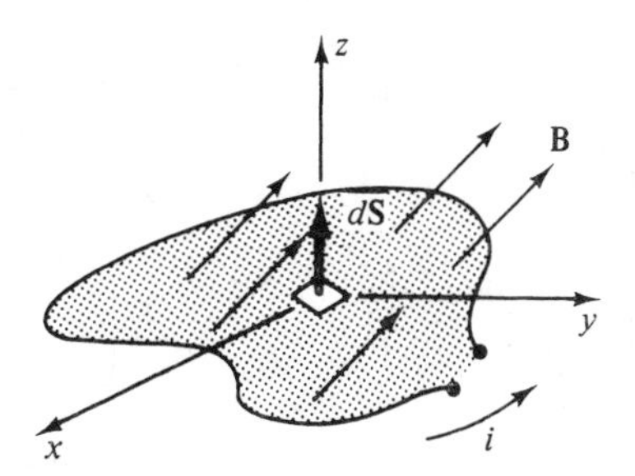

Figure 4-8 B flowing through area of a conducting loop.

The induced voltage is given by Faraday's law as

$$v = -\frac{d}{dt} \int_S \left(0.05 \cos 10^3 t \left(\frac{\mathbf{a}_y + \mathbf{a}_z}{\sqrt{2}} \right) \right) \bullet (dS\mathbf{a}_z)$$

$$= -\frac{d}{dt} \left(\frac{(0.05 \cos 10^3 t)(0.65)}{\sqrt{2}} \right) = 23.0 \sin 10^3 t \quad \text{V}$$

The field is decreasing in the first half cycle of the cosine function. The direction of i in a closed circuit must be such as to oppose this decrease. Thus, the conventional current must have the direction shown in Figure 4-8.

Solved Problem 4.3 In region *1* of Figure 4-9, $\mathbf{B}_1 = 1.2\mathbf{a}_x + 0.8\mathbf{a}_y + 0.4\mathbf{a}_z$ T. Find $\mathbf{H}_2$ (i.e., $\mathbf{H}$ at $z = 0^+$) and the angles between the field vectors and tangent to the interface.

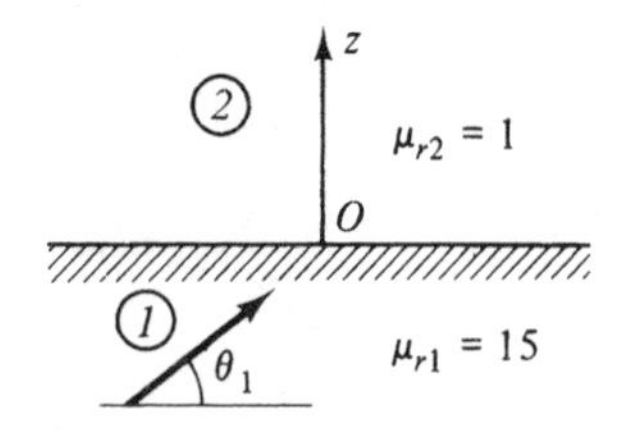

Figure 4-9 Magnetic field boundary problem.

Solution: Write $\mathbf{H}_1$ directly below $\mathbf{B}_1$. Then write those components of $\mathbf{H}_2$ and $\mathbf{B}_2$ which follow directly from the two rules **B** *normal is continuous* and **H** *tangential is continuous across a current-free interface.*

$$\begin{array}{lllll} \mathbf{B}_1 = 1.2\mathbf{a}_x & +0.8\mathbf{a}_y & +0.4\mathbf{a}_z & & \text{T} \\ \mathbf{H}_1 = \dfrac{1}{\mu_0}(8.0\mathbf{a}_x & +5.33\mathbf{a}_y & +2.67\mathbf{a}_z)10^{-2} & & \text{A/m} \\ \mathbf{H}_2 = \dfrac{1}{\mu_0}(8.0\mathbf{a}_x & +5.33\mathbf{a}_y & 10^2\mu_0 H_{2z})10^{-2} & & \text{A/m} \\ \mathbf{B}_1 = B_{2x}\mathbf{a}_x & +B_{2y}\mathbf{a}_y & +0.4\mathbf{a}_z & & \text{T} \end{array}$$

Now the remaining terms follow directly:

$$B_{2x} = \mu_0\mu_{r2}H_{2x} = 8.0\times10^{-2} \quad \text{T}$$

$$B_{2y} = \mu_0\mu_{r2}H_{2y} = 5.33\times10^{-2} \quad \text{T}$$

$$H_{2z} = \frac{B_{2z}}{\mu_0\mu_{r2}} = \frac{0.4}{\mu_0} \quad \text{A/m}$$

Angle θ_1 is $90° - \alpha_1$, where α_1 is the angle between $\mathbf{B}_1$ and the normal $\mathbf{a}_z$.

$$\cos\alpha_1 = \frac{\mathbf{B}_1 \bullet \mathbf{a}_z}{|\mathbf{B}_1|} = 0.27$$

Hence, $\alpha_1 = 74.5°$ and $\theta_1 = 15.5°$. Similarly, $\theta_2 = 76.5°$.

Solved Problem 4.4 A current sheet $\mathbf{K} = 6.5\mathbf{a}_z$ A/m at $x = 0$ separates region *1*, $x < 0$, where $\mathbf{H}_1 = 10\mathbf{a}_y$ A/m and region *2*, $x > 0$. Find $\mathbf{H}_2$ at $x = 0^+$.

Solution: Nothing is said about the permeabilities of the two regions. However, since $\mathbf{H}_1$ is entirely tangential, a change in permeability would have no effect ($B_{n1} = 0$, $B_{n2} = 0$ and therefore, $H_{n2} = 0$). Now,

$$
\begin{aligned}
&(\mathbf{H}_1 - \mathbf{H}_2)\times \mathbf{a}_{n12} = \mathbf{K} \\
&(10\mathbf{a}_y - H_{2y}\mathbf{a}_y)\times \mathbf{a}_x = 6.5\mathbf{a}_z \\
&(10 - H_{2y})(-\mathbf{a}_z) = 6.5\mathbf{a}_z \\
&H_{2y} = 16.5 \quad \mathrm{A/m}
\end{aligned}
$$

Thus, $\mathbf{H}_2 = 16.5\mathbf{a}_y \quad \mathrm{A/m}$.

Solved Problem 4.5 Given $\mathbf{H} = H_m\, e^{j(\omega t + \beta z)}\mathbf{a}_x$ in free space, find **E**.

Solution: Using Ampere's law (no current **J**)

$$
\nabla \times \mathbf{H} = \frac{\partial \mathbf{D}}{\partial t}
$$

$$
\frac{\partial}{\partial z}(H_m\, e^{j(\omega t + \beta z)})\mathbf{a}_y = \frac{\partial \mathbf{D}}{\partial t}
$$

$$
j\beta H_m\, e^{j(\omega t + \beta z)}\,\mathbf{a}_y = \frac{\partial \mathbf{D}}{\partial t}
$$

$$
\mathbf{D} = \frac{\beta H_m}{\omega}\, e^{j(\omega t + \beta z)}\,\mathbf{a}_y \quad \mathrm{C/m^2}
$$

and

$$
\mathbf{E} = \frac{\mathbf{D}}{\varepsilon_0} = \frac{\beta H_m}{\omega \varepsilon_0}\, e^{j(\omega t + \beta z)}\,\mathbf{a}_y \quad \mathrm{V/m}
$$

Chapter 5
ELECTROMAGNETIC WAVES

IN THIS CHAPTER:

- ✔ *Wave Equations and Solutions in Rectangular Coordinates*
- ✔ *Propagation in Various Media*
- ✔ *Interface Conditions for Normal Incidence*
- ✔ *Oblique Incidence and Snell's Laws*
- ✔ *Solved Problems*

This chapter deals with the solutions of Maxwell's equations for propagating electromagnetic waves such as radio waves. Since many regions of interest are free of charge, it will be assumed that the charge density $\rho = 0$. Moreover, linear isotropic materials will be assumed, with $\mathbf{D} = \varepsilon\mathbf{E}$, $\mathbf{B} = \mu\mathbf{H}$, and $\mathbf{J} = \sigma\mathbf{E}$.

Wave Equations and Solutions in Rectangular Coordinates

With the above assumptions and with time dependence $e^{j\omega t}$ for both **E** and **H**, Maxwell's equations (Table 4-1) become

$$\nabla \times \mathbf{H} = (\sigma + j\omega\varepsilon)\mathbf{E} \tag{1}$$

$$\nabla \times \mathbf{E} = -j\omega\mu\mathbf{H} \tag{2}$$

$$\nabla \bullet \mathbf{E} = 0 \tag{3}$$

$$\nabla \bullet B = 0 \tag{4}$$

Taking the curl of (1) and (2),

$$\nabla \times (\nabla \times \mathbf{H}) = (\sigma + j\omega\varepsilon)(\nabla \times \mathbf{E})$$
$$\nabla \times (\nabla \times \mathbf{E}) = -j\omega\mu(\nabla \times \mathbf{H})$$

Using the vector identity

$$\nabla \times (\nabla \times \mathbf{F}) = \nabla(\nabla \bullet \mathbf{F}) - \nabla^2\mathbf{F}$$

where the *Laplacian* of a vector **F** in *Cartesian coordinates* is

$$\nabla^2\mathbf{F} = (\nabla^2 F_x)\mathbf{a}_x + (\nabla^2 F_y)\mathbf{a}_y + (\nabla^2 F_z)\mathbf{a}_z$$

and, also in Cartesian coordinates,

$$\nabla^2 F_i = \frac{\partial^2 F_i}{\partial x^2} + \frac{\partial^2 F_i}{\partial y^2} + \frac{\partial^2 F_i}{\partial z^2}$$

Substituting the vector identity for the "double curls" and using (3) and (4) yield the *vector wave equations*

$$\nabla^2\mathbf{H} = j\omega\mu(\sigma + j\omega\varepsilon)\mathbf{H} = \gamma^2\mathbf{H}$$
$$\nabla^2\mathbf{E} = j\omega\mu(\sigma + j\omega\varepsilon)\mathbf{E} = \gamma^2\mathbf{E}$$

The *propagation constant* γ is the square root of γ^2 whose real and imaginary parts are

$$\gamma = \alpha + j\beta$$

with

$$\alpha = \omega\sqrt{\frac{\mu\varepsilon}{2}\left(\sqrt{1 + \left(\frac{\sigma}{\omega\varepsilon}\right)^2} - 1\right)} \tag{5}$$

$$\beta = \omega\sqrt{\frac{\mu\varepsilon}{2}\left(\sqrt{1+\left(\frac{\sigma}{\omega\varepsilon}\right)^2}+1\right)} \tag{6}$$

The familiar *scalar wave equation in one dimension*

$$\frac{\partial^2 F}{\partial z^2} = \frac{1}{u^2}\frac{\partial^2 F}{\partial t^2}$$

has solutions of the form $F = f(z - ut)$ and $F = g(z + ut)$, where f and g are arbitrary functions. These represent waves traveling with speed u in the $+z$ and $-z$ directions, respectively. In Figure 5-1, the first solutions is shown at $t = 0$ and $t = t_1$; the wave has advanced in the $+z$ direction at a distance of ut_1 in the time interval t_1.

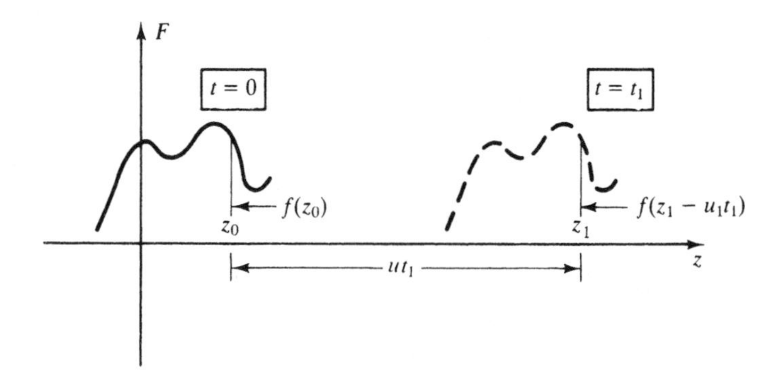

Figure 5-1 Wave propagating in the $+z$ direction.

For the particular choices,

$$f = Ce^{-j\omega z/u} \text{ and } g = De^{+j\omega z/u}$$

harmonic waves of angular frequency ω are obtained

$$F = Ce^{j(\omega t-\beta z)} \text{ and } F = De^{j(\omega t+\beta z)}$$

in which $\beta = \omega/u$. Of course, the real and imaginary parts are also solutions to the wave equation. One of these solutions,

$$F = C\sin(\omega t - \beta z)$$

is shown in Figure 5-2 at $t = 0$ and $t = \pi/2\omega$. In this interval, the wave has advanced in the $+z$ direction a distance $d = u(\pi/2\omega) = \pi/2\beta$.

At any fixed t, the waveform repeats itself when z changes by $2\pi/\beta$; the distance

$$\lambda = \frac{2\pi}{\beta} \tag{7}$$

is called the *wavelength.*

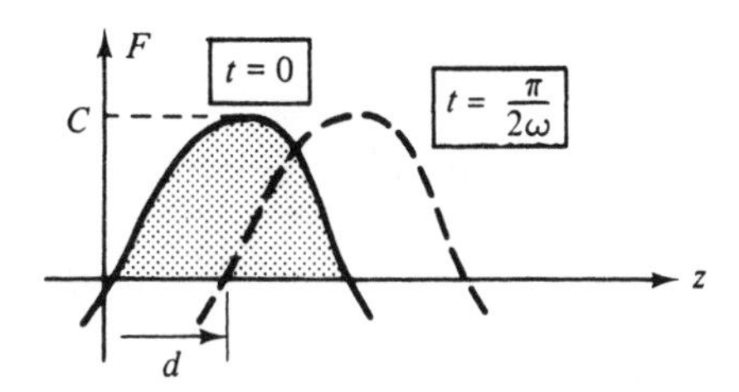

Figure 5-2 Harmonic wave propagation.

The wavelength and the *frequency* $f = \omega/2\pi$ are related by

$$\lambda = \frac{u}{f}$$

The vector wave equations have solutions similar to those just discussed. For Cartesian coordinates, the wave equation for **H** can be rewritten in the form

$$\frac{\partial^2 \mathbf{H}}{\partial x^2} + \frac{\partial^2 \mathbf{H}}{\partial y^2} + \frac{\partial^2 \mathbf{H}}{\partial z^2} = \gamma^2 \mathbf{H}$$

Of particular interest are solutions for *plane waves* that depend on only one spatial coordinate. Then the equation becomes

$$\frac{\partial^2 \mathbf{H}}{\partial z^2} = \gamma^2 \mathbf{H}$$

which, for an assumed time dependence $e^{j\omega t}$, is the vector analog of the one-dimensional scalar wave equation. Solutions are as above, in terms of the propagation constant γ.

$$\mathbf{H}(z,t) = H_0 \, e^{\pm \gamma z} \, e^{j\omega t} \, \mathbf{a}_H$$

The corresponding solutions for the electric field are

$$\mathbf{E}(z,t) = E_0 \, e^{\pm \gamma z} \, e^{j\omega t} \, \mathbf{a}_E$$

The fixed unit vectors $\mathbf{a}_H$ and $\mathbf{a}_E$ are orthogonal and neither field has a component in the direction of propagation. This being the case, one can rotate the axes to put one of the fields, say **E**, along the x-axis. Then, from Maxwell's equation (2) it follows that **H** will lie along the $\pm y$-axis for propagation in the $\pm z$-direction.

Propagation in Various Media

The plane wave solutions obtained above depend on the properties μ, ε, and σ of the medium due to their relation to γ. For a region in which there is some conductivity but not much (e.g., moist earth, sea water), the solution to the wave equation in **E** is taken to be

$$\mathbf{E} = E_0 \, e^{-\gamma z} \mathbf{a}_x$$

where the $e^{j\omega t}$ time dependence has been suppressed. Then, from (2),

$$\mathbf{H} = \sqrt{\frac{\sigma + j\omega\varepsilon}{j\omega\mu}} \, E_0 \, e^{-\gamma z} \mathbf{a}_y$$

The ratio E/H is characteristic of the medium for plane waves.

For the above **E** and **H**, the *intrinsic impedance of the medium* is defined by

$$\eta = \frac{E_x}{H_y} = \sqrt{\frac{j\omega\mu}{\sigma + j\omega\varepsilon}} = |\eta| e^{j\theta} \ (\Omega) \qquad (8)$$

where $\tan 2\theta = \sigma/\omega\varepsilon$. If the wave propagates in the $-z$-direction, $E_x/H_y = -\eta$.

Inserting the time factor $e^{j\omega t}$ and writing $\gamma = \alpha + j\beta$ results in the following equations for the field in the partially conducting region:

$$\mathbf{E}(z,t) = E_0\, e^{-\alpha z}\, e^{j(\omega t-\beta z)}\mathbf{a}_x$$
$$\mathbf{H}(z,t) = \frac{E_0}{|\eta|} e^{-\alpha z}\, e^{j(\omega t-\beta z-\theta)}\mathbf{a}_y$$

The factor $e^{-\alpha z}$ attenuates the magnitudes of both **E** and **H** as they propagate in the $+z$ direction.

For perfect dielectrics, $\sigma = 0$ and so

$$\alpha = 0, \quad \beta = \omega\sqrt{\mu\varepsilon}, \quad \eta = \sqrt{\frac{\mu}{\varepsilon}}, \theta = 0° \tag{9}$$

Since $\alpha = 0$, there is no attenuation of the **E** and **H** waves. The waves are

$$\mathbf{E}(z,t) = E_0\, e^{j(\omega t-\beta z)}\mathbf{a}_x$$
$$\mathbf{H}(z,t) = \frac{E_0}{\eta}\, e^{j(\omega t-\beta z)}\mathbf{a}_y$$

The velocity of propagation is $u = \dfrac{\omega}{\beta} = \dfrac{1}{\sqrt{\mu\varepsilon}}$. In free space, the permeability and permittivity, respectively, are $\mu_0 = 4\pi \times 10^{-7}$ H/m and $\varepsilon_0 = 10^{-9}/36\pi$ F/m, giving $u = 3 \times 10^8$ m/s and $\eta = 120\pi\ \Omega$.

Materials are ordinarily classified as *good conductors* if $\sigma >> \omega\varepsilon$ in the range of practical frequencies. Therefore, the propagation constant and the intrinsic impedance are

$$\gamma = \alpha + j\beta, \quad \alpha = \beta \approx \sqrt{\frac{\omega\mu\sigma}{2}} = \sqrt{\pi f \mu\sigma}, \quad \eta \approx \sqrt{\frac{\omega\mu}{\sigma}}\angle 45°$$

It is seen that for all conductors the waves are attenuated.

The depth at which the fields are attenuated to 37% of their original value is given by the *skin depth* δ

$$\delta = \frac{1}{\alpha} = \frac{1}{\sqrt{\pi f \mu\sigma}} \text{ m}$$

Example 5.1 Assume a field $\mathbf{E}(z,t) = 1.0e^{-\alpha z}e^{j(\omega t-\beta z)}\mathbf{a}_x$ V/m with frequency $f = 100$ MHz at the surface of a copper conductor, $\sigma = 5.8 \times 10^7$ S/m, located at $z > 0$, as shown in Figure 5-3. Examine the attenuation as the wave propagates into the conductor.

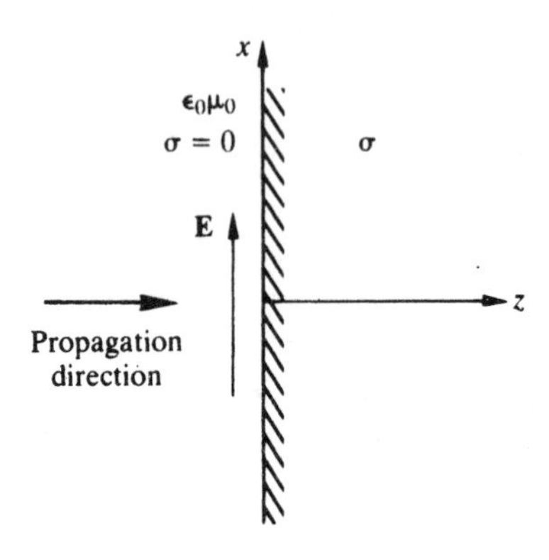

Figure 5-3 Plane wave propagating into copper.

Solution: At depth z, the magnitude of the field is

$$|\mathbf{E}| = 1.0e^{-\alpha z} = 1.0e^{-z/\delta}$$

where

$$\delta = \frac{1}{\sqrt{\pi f \mu \sigma}} = 6.61 \quad \mu\text{m}$$

Therefore, after just 6.61 μm the field amplitude is attenuated to $e^{-1} = 37\%$ of its initial value. At $z = 5\delta$ or 33 μm, the magnitude is 0.67 % of its initial value—practically zero.

Interface Conditions for Normal Incidence

When a traveling wave reaches an interface between two different regions, it is partly reflected and partly transmitted, with the magnitudes of the two parts determined by the constants of the two regions. In Figure 5-4, a traveling wave approaches the interface $z = 0$ from region 1

($z < 0$). The incident wave $\mathbf{E}^i$ and reflected wave $\mathbf{E}^r$ will be evaluated at $z = 0^-$. The transmitted wave $\mathbf{E}^t$ is in region 2 ($z > 0$) and will be evaluated at the interface at $z = 0^+$. Normal incidence is assumed (i.e., the wave is traveling in a direction perpendicular to the interface).

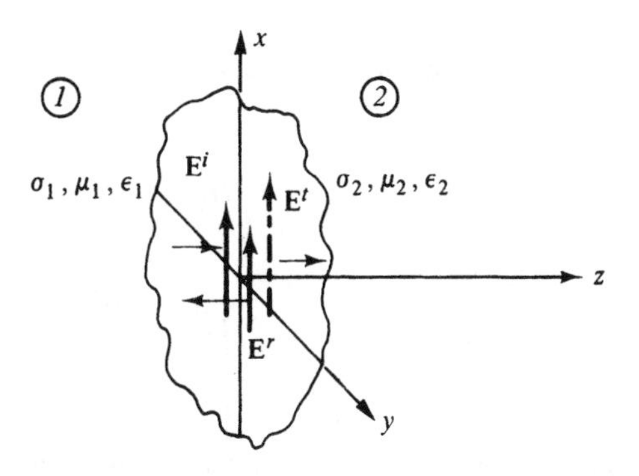

Figure 5-4 Normal wave incidence upon a planar boundary.

The equations for **E** and **H** can be written as

$$\mathbf{E}^i(z,t) = E_0^i\, e^{-\gamma_1 z}\, e^{j\omega t}\mathbf{a}_x$$
$$\mathbf{E}^r(z,t) = E_0^r\, e^{\gamma_1 z}\, e^{j\omega t}\mathbf{a}_x$$
$$\mathbf{E}^t(z,t) = E_0^t\, e^{-\gamma_2 z}\, e^{j\omega t}\mathbf{a}_x$$
$$\mathbf{H}^i(z,t) = H_0^i\, e^{-\gamma_1 z}\, e^{j\omega t}\mathbf{a}_y$$
$$\mathbf{H}^r(z,t) = H_0^r\, e^{\gamma_1 z}\, e^{j\omega t}\mathbf{a}_y$$
$$\mathbf{H}^t(z,t) = H_0^t\, e^{-\gamma_2 z}\, e^{j\omega t}\mathbf{a}_y$$

One of the six constants—it is almost always E_0^i—may be taken as real. Under the interface conditions about to be derived, one or more of the remaining five may turn out to be complex.

With normal incidence, **E** and **H** are entirely tangential to the interface and are, according to the boundary conditions (Chapter 4), continuous. At $z = 0$, this implies

$$E_0^i + E_0^r = E_0^t, \qquad H_0^i - H_0^r = H_0^t$$

Furthermore, the ratios for intrinsic impedance for the waves are equal to $\pm E_x/H_y$ and thus are

$$\frac{E_0^i}{H_0^i} = \eta_1, \quad \frac{E_0^r}{H_0^r} = -\eta_1, \quad \frac{E_0^t}{H_0^t} = \eta_2$$

Using the above five equations, the following ratios can be derived:

$$\Gamma_E = \frac{E_0^r}{E_0^i} = \frac{\eta_2 - \eta_1}{\eta_2 + \eta_1}, \quad \Gamma_H = \frac{H_0^r}{H_0^i} = \frac{\eta_1 - \eta_2}{\eta_2 + \eta_1} \tag{10}$$

$$\mathrm{T}_E = \frac{E_0^t}{E_0^i} = \frac{2\eta_2}{\eta_2 + \eta_1}, \quad \mathrm{T}_H = \frac{H_0^t}{H_0^i} = \frac{2\eta_1}{\eta_2 + \eta_1} \tag{11}$$

where the Γs and Ts represent *reflection coefficients* and *transmission coefficients*, respectively. The intrinsic impedances are defined in (8) and (9).

Example 5.2 Traveling **E** and **H** waves in free space (region 1) are normally incident upon a planar interface with a perfect dielectric (region 2, $\sigma_2 = 0$) for which $\varepsilon_{r2} = 3.0$. Compare the magnitudes of the incident, reflected and transmitted **E** and **H** waves at the interface.

Solution: The magnitudes are related by the reflection and transmission coefficients from (10) and (11).

$$\eta_1 = \eta_0 = 120\pi\ \Omega, \qquad \eta_2 = \sqrt{\frac{\mu}{\varepsilon}} = \frac{120\pi}{\sqrt{\varepsilon_{r2}}} = 217.7\ \Omega$$

$$\frac{E_0^r}{E_0^i} = \frac{\eta_2 - \eta_1}{\eta_2 + \eta_1} = -0.268, \quad \frac{H_0^r}{H_0^i} = \frac{\eta_1 - \eta_2}{\eta_2 + \eta_1} = 0.268$$

$$\frac{E_0^t}{E_0^i} = \frac{2\eta_2}{\eta_2 + \eta_1} = 0.732, \quad \frac{H_0^t}{H_0^i} = \frac{2\eta_1}{\eta_2 + \eta_1} = 1.268$$

When waves traveling in a perfect dielectric ($\sigma_1 = 0$) are normally incident with a perfect conductor ($\sigma_2 = \infty$, $\eta_2 = 0$), the reflected wave in combination with the incident wave produces a *standing wave.* In such a wave, which is readily demonstrated on a clamped taut string, the oscillations at all points of a half-wavelength interval are in time phase. The combination of incident and reflected waves may be written

$$\mathbf{E}(z,t) = [E_0^i\, e^{j(\omega t - \beta z)} + E_0^r e^{j(\omega t + \beta z)}]\mathbf{a}_x = e^{j\omega t}(E_0^i\, e^{-j\beta z} + E_0^r e^{j\beta z})\mathbf{a}_x$$

Since $\eta_2 = 0$, $E_0^r / E_0^i = -1$, and

$$\mathbf{E}(z,t) = e^{j\omega t}(E_0^i\, e^{-j\beta z} - E_0^i e^{j\beta z})\mathbf{a}_x = -2jE_0^i \sin\beta z\, e^{j\omega t}\, \mathbf{a}_x$$

or, taking the real part,

$$\mathbf{E}(z,t) = 2E_0^i \sin\beta z \sin\omega t\ \mathbf{a}_x$$

The standing wave is shown in Figure 5-5 at time intervals of $T/8$, where $T = 2\pi/\omega$ is the period. At $t = 0$, $\mathbf{E} = 0$ everywhere; at $t = 1(T/8)$, the endpoints of the **E** vectors lie on the sine curve *1*; at $t = 2(T/8)$, the endpoints of the **E** vectors lie on the sine curve *2*; and so forth. Sine curves *2* and *6* form an envelope for the oscillations; the amplitude of this envelope is twice the amplitude of the incident wave. Note that the adjacent half-wavelength segments are 180° out of phase with each other.

Oblique Incidence and Snell's Laws

An incident wave that approaches a plane interface between two different media generally will result in a transmitted wave in the second medium and a reflected wave in the first. The *plane of incidence* is the plane containing the incident wave normal and the local normal to the interface. In Figure 5-6, the plane of incidence is the *xz* plane. The normals to the reflected and transmitted waves also lie in the plane of incidence.

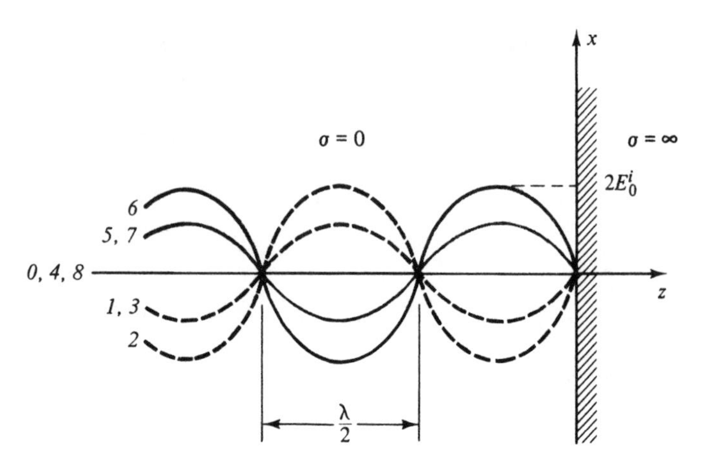

Figure 5-5 Standing wave for normally incident wave.

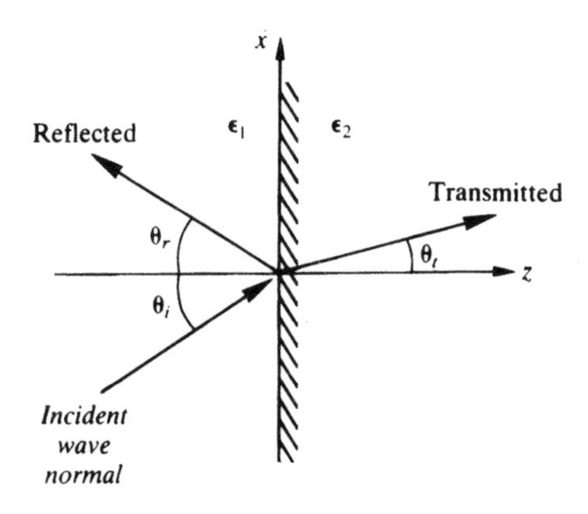

Figure 5-6 Oblique wave incidence problem.

The *angle of incidence* θ_i, *the angle of reflection* θ_r, *and the angle of transmission* θ_t—all defined in Figure 5-6—obey *Snell's law of reflection,*

$$\theta_i = \theta_r \tag{12}$$

and *Snell's law of refraction,*

$$\frac{\sin\theta_i}{\sin\theta_t} = \sqrt{\frac{\mu_2\varepsilon_2}{\mu_1\varepsilon_1}} \tag{13}$$

Example 5.3 A wave is incident at an angle of 30° from air to teflon, $\varepsilon_r = 2.1$. Calculate the angle of transmission and repeat with an interchange of the regions. The permeability of both regions is μ_0.

Solution: Using Snell's law of refraction (13),

$$\frac{\sin\theta_i}{\sin\theta_t} = \frac{\sin 30^\circ}{\sin\theta_t} = \sqrt{\frac{\varepsilon_{r2}}{\varepsilon_{r1}}} = \sqrt{2.1} \quad \text{or} \quad \theta_t = 20.18^\circ$$

From teflon to air,

$$\frac{\sin\theta_i}{\sin\theta_t} = \frac{\sin 30^\circ}{\sin\theta_t} = \sqrt{\frac{\varepsilon_{r2}}{\varepsilon_{r1}}} = \frac{1}{\sqrt{2.1}} \quad \text{or} \quad \theta_t = 46.43^\circ$$

Supposing both media are of the same permeability, propagation from the optically denser medium ($\varepsilon_1 > \varepsilon_2$) results in $\theta_t > \theta_i$. As θ_i increases, an angle of incidence will be reached that results in $\theta_t = 90°$. At this *critical angle of incidence*, instead of a wave being transmitted into the second medium there will be a wave that propagates along the surface.

The *critical angle* is given by

$$\theta_c = \sin^{-1}\sqrt{\frac{\varepsilon_{r2}}{\varepsilon_{r1}}} \tag{14}$$

Example 5.4 What is the critical angle for a wave propagating from teflon into free space?

Solution: Using (14),

$$\theta_c = \sin^{-1}\frac{1}{\sqrt{2.1}} = 43.64^\circ$$

For an obliquely incident **E** that has its field vector parallel to the plane of incidence, an angle can be found such that there is no reflected wave.

The *Brewster angle* for which there is no reflected wave is

$$\theta_B = \tan^{-1}\sqrt{\frac{\varepsilon_{r2}}{\varepsilon_{r1}}} \tag{15}$$

Example 5.5 What is the Brewster angle for a wave traveling from air into glass for which $\varepsilon_r = 5.0$?

Solution: Using (15),

$$\theta_B = \tan^{-1}\sqrt{5.0} = 65.91°$$

Important Things to Remember:

- ✔ Plane waves have both **E** and **H** vectors perpendicular to the direction of propagation.
- ✔ In a lossy media ($\sigma \neq 0$), the skin depth δ is the distance where the plane wave amplitude is reduced to 37% of its original value.
- ✔ In general, for two media, an incident wave gives rise to a reflected wave and a transmitted wave.
- ✔ For oblique incidence problems, Snell's laws govern the angles of reflection and transmission.

Solved Problems

Solved Problem 5.1 In free space, $\mathbf{E} = 10^3 \sin(\omega t - \beta z)\mathbf{a}_y$ V/m. Obtain $\mathbf{H}(z,t)$.

Solution: Examination of the phase, $\omega t - \beta z$, shows that the direction of propagation is $+z$. Since $\mathbf{E} \times \mathbf{H}$ must also be in the $+z$-direction, $\mathbf{H}$ must have direction $-\mathbf{a}_x$. Consequently,

$$\frac{E_y}{-H_x} = \eta_0 = 120\pi\ \Omega \quad \text{or} \quad H_x = -\frac{10^3}{120\pi}\sin(\omega t - \beta z)\ \text{A/m}$$

and

$$\mathbf{H} = -\frac{10^3}{120\pi}\sin(\omega t - \beta z)\mathbf{a}_x\ \text{A/m}$$

Solved Problem 5.2 Calculate the intrinsic impedance η, the propagation constant γ and the wave velocity u for a conducting medium in which $\sigma = 5.8 \times 10^7$ S/m, $\mu_r = 1$ and the frequency is $f = 100$ MHz.

Solution: For a good conductor,

$$\eta = \sqrt{\frac{\omega\mu}{\sigma}}\angle 45° = 3.69 \times 10^{-3}\angle 45°\ \Omega$$

$$\gamma = \sqrt{\omega\mu\sigma}\angle 45° = 2.14 \times 10^5\angle 45°\ \text{m}^{-1}$$

$$\alpha = \beta = 1.51 \times 10^5, \quad \delta = \frac{1}{\alpha} = 6.61\ \mu\text{m}, \quad u = \omega\delta = 4.15 \times 10^3\ \text{m/s}$$

Solved Problem 5.3 A plane wave traveling in the $+z$-direction in free space ($z < 0$) is normally incident on a conductor ($z > 0$) for which $\sigma = 6.17 \times 10^7$, $\mu_r = 1$ and the frequency is $f = 1.5$ MHz. The free space **E** wave has an amplitude of 1.0 V/m and at the interface, the total **E** in free space is given by

$$\mathbf{E}(0,t) = 1.0\sin 2\pi ft\,\mathbf{a}_y \quad \text{V/m}$$

Find **H**(z,t) for $z > 0$.

Solution: For $z > 0$, in complex form,

$$\mathbf{E}(z,t) = 1.0e^{-\alpha z}e^{j(2\pi ft - \beta z)}\mathbf{a}_y \quad \text{V/m}$$

where the imaginary part will ultimately be taken. In the conductor,

$$\alpha = \beta = \sqrt{\pi f\mu\sigma} = \sqrt{\pi(1.5 \times 10^6)(4\pi \times 10^{-7})(61.7 \times 10^6)} = 1.91 \times 10^4$$

$$\eta = \sqrt{\frac{\omega\mu}{\sigma}}\angle 45° = 4.38 \times 10^{-4}\,e^{j\pi/4}\ \Omega$$

Then, since $E_y/(-H_x) = \eta$,

$$\mathbf{H}(z,t) = -2.28 \times 10^3 \, e^{-\alpha z} \, e^{j(2\pi ft - \beta z - \pi/4)} \mathbf{a}_x \quad \text{A/m}$$

or, taking the imaginary part,

$$\mathbf{H}(z,t) = -2.28 \times 10^3 \, e^{-\alpha z} \sin(2\pi ft - \beta z - \pi / 4) \mathbf{a}_x \quad \text{A/m}$$

Solved Problem 5.4 Determine the amplitudes of the reflected and transmitted **E** and **H** at the interface shown in Figure 5-7 if $E_0^i = 1.5 \times 10^{-3}$ in region *1* in which $\varepsilon_{r1} = 8.5$, $\mu_1 = 1$ and $\sigma_1 = 0$. Region *2* is free space. Assume normal incidence.

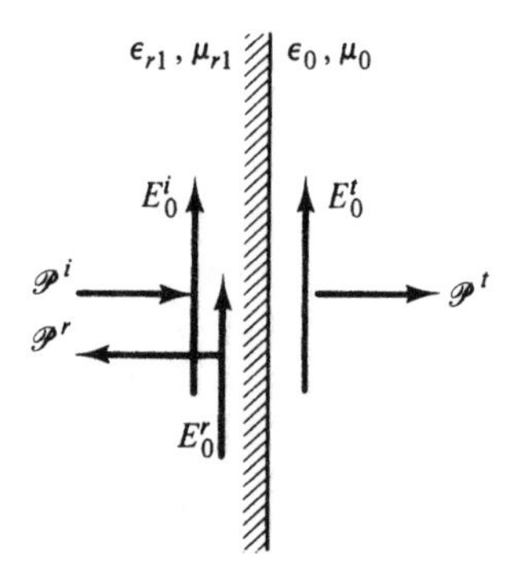

Figure 5-7 Normal incidence of a plane wave.

Solution: Computing the reflection and transmission coefficients,

$$\eta_1 = \sqrt{\frac{\mu_{r1}\mu_0}{\varepsilon_{r1}\varepsilon_0}} = 129\ \Omega, \qquad \eta_2 = 120\pi = 377\ \Omega$$

$$E_0^r = \frac{\eta_2 - \eta_1}{\eta_2 + \eta_1} E_0^i = 7.35 \times 10^{-4}\ \text{V/m}$$

$$E_0^t = \frac{2\eta_2}{\eta_2 + \eta_1} E_0^i = 2.24 \times 10^{-3}\ \text{V/m}$$

$$H_0^i = \frac{E_0^i}{\eta_1} = 1.16 \times 10^{-5}\ \text{A/m}$$

$$H_0^r = \frac{\eta_1 - \eta_2}{\eta_2 + \eta_1} H_0^i = -5.69 \times 10^{-6}\ \text{A/m}$$

$$H_0^t = \frac{2\eta_1}{\eta_2 + \eta_1} H_0^i = 5.91 \times 10^{-6}\ \text{A/m}$$

Chapter 6
TRANSMISSION LINES

IN THIS CHAPTER:

- ✔ *Distributed Parameters and Transmission Line Models*
- ✔ *Sinusoidal Steady State Excitation*
- ✔ *Smith Chart Analysis*
- ✔ *Transients in Lossless Lines*
- ✔ *Solved Problems*

Unguided propagation of electromagnetic waves was investigated in Chapter 5. In this chapter, the transmission of energy will be studied when the waves are guided by two conductors in a dielectric medium. Exact analysis of this two-conductor *transmission line* requires field theory. However, the performance of the system can be predicted by modeling the transmission line with distributed parameters and using voltages and currents associated with the electric and magnetic fields. For the problems in this chapter, it is assumed that the parameters are constant along the length of the line.

Distributed Parameters and Transmission Line Models

The incremental distributed parameters per unit length of line are inductance, capacitance, resistance of the conductors, and conductance of the dielectric medium. In general, the parameters depend on the geometry of the configuration, material properties, and in some cases, the frequency. In the following summary list, the dependence on geometry is represented by a *geometrical factor* "GF."

Capacitance: $C = \pi\varepsilon_d(\mathrm{GF_C})$ F/m [ε_d = permittivity of dielectric]

Conductance: $G = \dfrac{C}{\varepsilon_d}\sigma_d$ S/m [σ_d = conductivity of dielectric]

Inductance (external):

$$L_e = \frac{\mu_d}{\pi}(\mathrm{GF_L}) \quad \mathrm{H/m} \quad [\mu_d = \text{permeability of dielectric}]$$

DC Resistance (useful up to approximately 10 kHz):

$$R_{dc} = \frac{1}{\sigma_c \pi}(\mathrm{GF_{RDC}}) \quad \Omega/\mathrm{m} \quad [\sigma_c = \text{conductivity of conductors}]$$

AC Resistance (useful above approximately 10 kHz):

$$R_{ac} = \frac{1}{2\pi\sigma_c\delta}(\mathrm{GF_{RAC}}) \quad \Omega/\mathrm{m} \quad \left[\delta = \frac{2}{\sqrt{\pi f \mu_c \sigma_c}} \equiv \text{skin depth}\right]$$

Inductance (internal):

$$L_i = \begin{cases} R_{ac}/2\pi f & \mathrm{H/m} \quad \text{for} \quad f > 10\ \mathrm{kHz} \\ \mu_0/4\pi & \mathrm{H/m} \quad \text{for} \quad f < 10\ \mathrm{kHz} \end{cases}$$

Total inductance: $L_t = L_e + L_i \approx L_e$ for many practical problems

For three common line configurations, the geometrical factors are as follows:

Coaxial Cable (inner radius a, outer radius b, outer thickness of shield t):

$$\mathrm{GF_C} = \frac{2}{\ln(b/a)}, \qquad \mathrm{GF_L} = \frac{1}{\mathrm{GF_C}}$$

$$\mathrm{GF_{RDC}} = \frac{1}{a^2} + \frac{1}{t(b+t)}, \qquad \mathrm{GF_{RAC}} = \frac{1}{a} + \frac{1}{b} \quad \text{for} \quad t \gg \delta$$

Parallel wire transmission line (wire radius a, separation d):

$$\mathrm{GF_C} = \frac{1}{\mathrm{GF_L}}, \qquad \mathrm{GF_L} = \cosh^{-1}\frac{d}{2a} \approx \ln\frac{d}{a} \quad \text{for} \quad d >> a$$

$$\mathrm{GF_{RDC}} = \frac{2}{a^2}, \qquad \mathrm{GF_{RAC}} = \frac{2}{a}$$

Parallel plates (width w, thickness t, separation d):

$$\mathrm{GF_C} = \frac{w}{\pi d}, \qquad \mathrm{GF_L} = \frac{1}{\mathrm{GF_C}}$$

$$\mathrm{GF_{RDC}} = \frac{2\pi}{wt}, \qquad \mathrm{GF_{RAC}} = \frac{4\pi}{w} \quad \text{for} \quad t >> \delta$$

The model in Figure 6-1, where R, L, G, and C are as given above, permits analysis of the line using voltages and currents. For within a cell of length Δx, the voltages across the line at points a and b differ by

$$\Delta v(x,t) = (R\Delta x)\,i(x,t) + (L\Delta x)\frac{\partial i(x,t)}{\partial t}$$

In the limit as Δx goes to zero, this becomes

$$\frac{\partial v(x,t)}{\partial x} = R\,i(x,t) + L\frac{\partial i(x,t)}{\partial t} \tag{1}$$

Likewise, the current at point c differs from that at b by

$$\Delta i(x,t) = (G\Delta x)\,v(x,t) + (C\Delta x)\frac{\partial v(x,t)}{\partial t}$$

which, in the limit, gives

$$\frac{\partial i(x,t)}{\partial x} = G\,v(x,t) + C\frac{\partial v(x,t)}{\partial t} \tag{2}$$

Equations (1) and (2) are said to be *transmission line equations*. From (1) and (2), a set of second order equations can be derived for which

$$\frac{\partial^2 f(x,t)}{\partial x^2} = RG\,f(x,t) + (RC + LG)\frac{\partial f(x,t)}{\partial t} + LC\frac{\partial^2 f(x,t)}{\partial t^2} \tag{3}$$

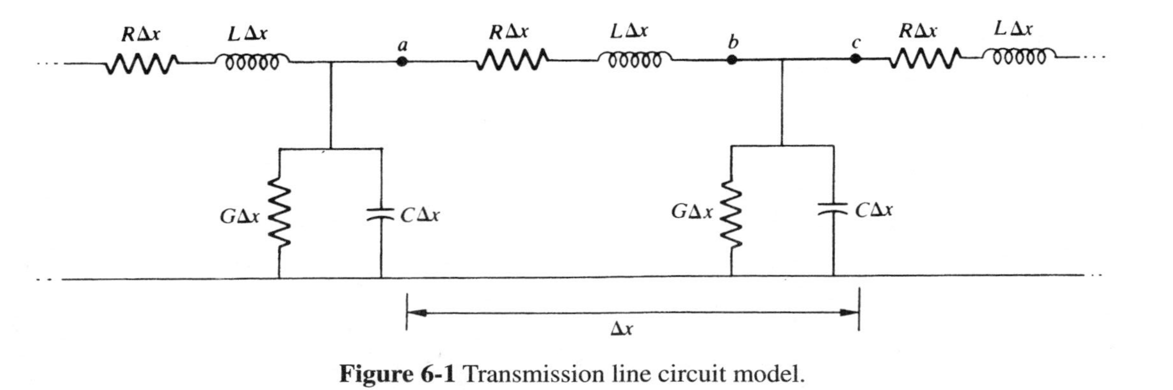

Figure 6-1 Transmission line circuit model.

for $f(x,t)$ = either $i(x,t)$ or $v(x,t)$. Now, (3) is an equation of hyperbolic type, similar to a wave equation. For a lossless line ($R = G = 0$), (3) is the one-dimensional scalar wave equation discussed in Chapter 5. Thus, it is known in advance that transmission lines support voltage- and current-waves which can be reflected and/or transmitted at discontinuities (sites of abrupt parameter changes) in the line.

Sinusoidal Steady State Excitation

When the transmission line of Figure 6-1 is driven for a long time by a sinusoidal source (angular frequency ω), the voltage and current also become sinusoidal, with the same frequency:

$$v(x,t) = \text{Re}[\hat{V}(x)e^{j\omega t}], \qquad i(x,t) = \text{Re}[\hat{I}(x)e^{j\omega t}]$$

Here the phasors $\hat{V}(x)$ and $\hat{I}(x)$ are generally complex valued; often they are indicated in polar form (with the x-dependence suppressed) as

$$\hat{V} = \left|\hat{V}\right| \angle\phi_V, \qquad \hat{I} = \left|\hat{I}\right| \angle\phi_I$$

where ϕ denotes the angle between the complex phasor and the real axis. Steady state transmission line analysis is simplified when voltages and currents are replaced by their phasor representations.

Figure 6-2 models in the phasor domain a uniform line of length ℓ that is terminated in a (complex) load Z_R at the receiving end and is driven at the sending end by a generator with internal impedance Z_g and voltage $\hat{V}_g = V_{gm} \angle\theta$. The per-unit-length series impedance and shunt admittance of the line are given by

$$Z = R + j\omega L, \qquad Y = G + j\omega C$$

Note!

Distance from the receiving end is measured by the variable *x*; from the sending end by *d*. ***This convention is used throughout this chapter.***

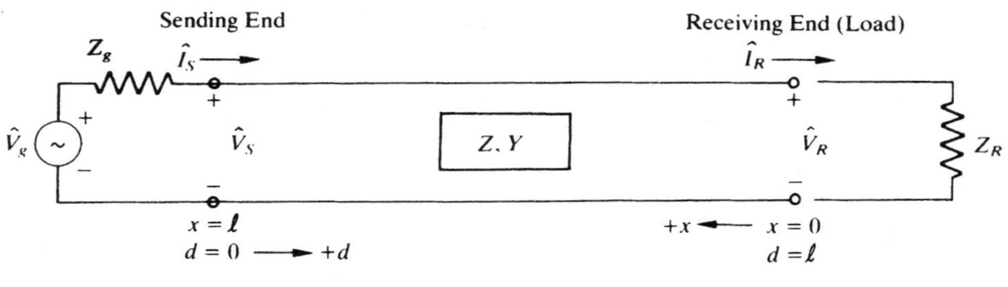

Figure 6-2 Sinusoidal excitation of transmission line.

Using phasor analysis, equations (1), (2), and (3) become ODEs in $\hat{V}$ and $\hat{I}$.

$$\frac{d\hat{V}(x)}{dx} = Z\,\hat{I}(x) \tag{4}$$

$$\frac{d\hat{I}(x)}{dx} = Y\,\hat{V}(x) \tag{5}$$

$$\frac{d^2\hat{F}(x)}{dx^2} = \gamma^2\,\hat{F}(x) \tag{6}$$

with $\gamma = \sqrt{ZY} = \alpha + j\beta$, the square root being chosen to make α and β nonnegative. Equation (6) is identical in form to the equation of plane waves (Chapter 5). It has traveling wave solutions (for x measured as the distance from the load)

$$\hat{V}(x) = \hat{V}^+ e^{\gamma x} + \hat{V}^- e^{-\gamma x} \equiv \hat{V}_{inc}(x) + \hat{V}_{refl}(x)$$

$$\hat{I}(x) = \hat{I}^+ e^{\gamma x} + \hat{I}^- e^{-\gamma x} \equiv \hat{I}_{inc}(x) + \hat{I}_{refl}(x)$$

The coefficients $\hat{V}^+$, etc., are phasors independent of x that are interrelated by the *characteristic impedance* Z_0 *and the boundary reflection coefficient* Γ_R, defined as

$$Z_0 \equiv \frac{\hat{V}^+}{\hat{I}^+} = -\frac{\hat{V}^-}{\hat{I}^-} = \sqrt{\frac{Z}{Y}}$$

$$\Gamma_R \equiv \frac{\hat{V}^-}{\hat{V}^+} = -\frac{\hat{I}^-}{\hat{I}^+} = \frac{\hat{V}_{refl}(0)}{\hat{V}_{inc}(0)}$$

It is straightforward to express Γ_R in terms of characteristic and load impedances:

$$\Gamma_R = \frac{Z_R - Z_0}{Z_R + Z_0}$$

Then, if a pointwise reflection coefficient is defined by

$$\Gamma(x) = \frac{\hat{V}_{refl}(x)}{\hat{V}_{inc}(x)}$$

it follows that

$$\Gamma(x) = \Gamma_R e^{-2\gamma x} = \frac{Z_R - Z_0}{Z_R + Z_0} e^{-2\gamma x}$$

Similarly, if $Z(x) = \hat{V}(x)/\hat{I}(x)$ is the pointwise impedance looking back to the receiving end ($x = 0$), then

$$Z(x) = Z_0 \frac{1 + \Gamma(x)}{1 - \Gamma(x)}$$

The conditions at the sending end (again, using x as the distance variable measured from the receiving end) are

$$Z_S \equiv Z(\ell) = Z_0 \frac{1 + \Gamma(\ell)}{1 - \Gamma(\ell)}$$

$$\hat{V}_S = \hat{V}_g \frac{Z_S}{Z_S + Z_g}$$

$$\hat{I}_S = \frac{\hat{V}_S}{Z_S}$$

The average power received by the load and average power supplied at the sending end are calculated as

$$P_R = \tfrac{1}{2}\mathrm{Re}\left(\hat{V}_R \hat{I}_R^*\right) = \tfrac{1}{2}\left|\hat{I}_R\right|^2 \mathrm{Re}(Z_R)$$
$$= P_{inc}(x = 0) - P_{refl}(x = 0)$$

$$P_S = \tfrac{1}{2}\mathrm{Re}\left(\hat{V}_S \hat{I}_S^*\right) = \tfrac{1}{2}\left|\hat{I}_S\right|^2 \mathrm{Re}(Z_S)$$

For frequencies such that $R << \omega L$ and $G << \omega C$ (e.g., above 1 MHz),

$$Z_0 = \sqrt{\frac{R + j\omega L}{G + j\omega C}} \approx \sqrt{\frac{L}{C}} = R_0$$

$$\gamma = \sqrt{(R + j\omega L)(G + j\omega C)} \approx \left(\frac{R}{2R_0} + \frac{GR_0}{2}\right) + j\omega\sqrt{LC} = \alpha + j\beta$$

$$u \approx \frac{1}{\sqrt{LC}} \quad \text{and} \quad \lambda = \frac{2\pi}{\beta} \approx \frac{1}{f\sqrt{LC}}$$

For the ideal lossless line ($R = G = 0$), the reflection coefficient is of constant magnitude

$$\Gamma(x) = \Gamma_R e^{-j2\beta x} = \left|\frac{Z_R - R_0}{Z_R + R_0}\right| \angle(\phi_R - 2\beta x)$$

where ϕ_R is the angle of Γ_R. The voltage is given by

$$\hat{V}(x) = \hat{V}^+[1 + \Gamma_R \angle(-2\beta x)]$$

which implies

$$\left|\hat{V}\right|_{\max} = \left|\hat{V}^+\right|(1 + |\Gamma_R|), \quad \left|\hat{V}\right|_{\min} = \left|\hat{V}^+\right|(1 - |\Gamma_R|)$$

Adjacent maxima and minima are separated by $\beta x = 90^\circ$ or one-quarter wavelength.

You Need to Know 

For the resulting wave, the *voltage standing wave ratio, VSWR*, is defined as

$$VSWR = \frac{\left|\hat{V}\right|_{\max}}{\left|\hat{V}\right|_{\min}} = \frac{1 + |\Gamma_R|}{1 - |\Gamma_R|}$$

Smith Chart Analysis

The Smith chart (Figure 6-3) is a graphical aid in solving high-frequency transmission line problems. The chart is essentially a polar plot of the reflection coefficient in terms of the normalized impedance $z(x)$:

$$z(x) \equiv \frac{Z(x)}{R_0} = r(x) + j\chi(x) = \frac{1 + \Gamma(x)}{1 - \Gamma(x)}$$

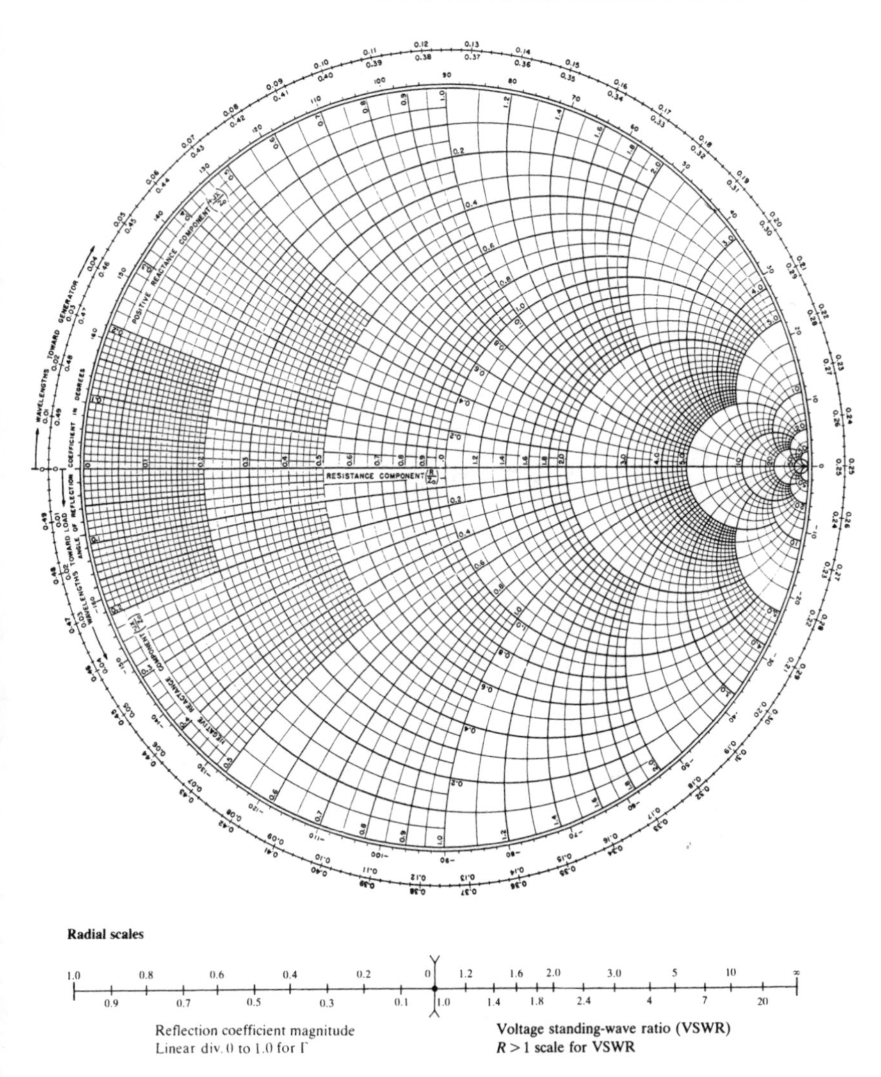

Figure 6-3 Smith chart.

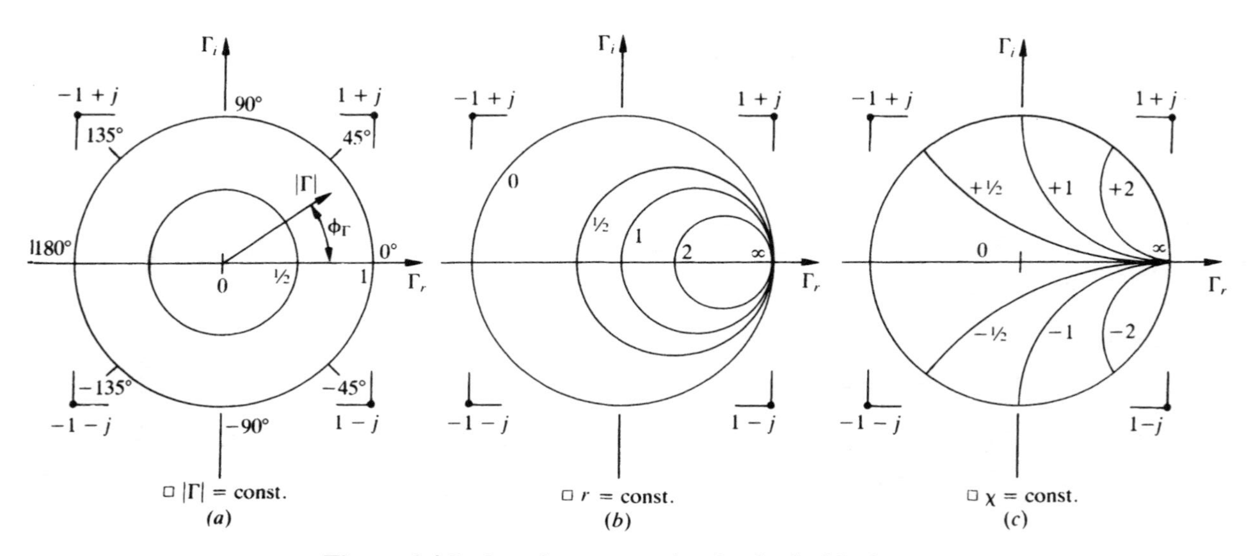

Figure 6-4 Scales of constant value for the Smith chart.

$$\Gamma(x) = \Gamma_R \angle(-2\beta x) = \left|\frac{r_0 + j\chi_0 - 1}{r_0 + j\chi_0 + 1}\right| \angle(\phi_R - 2\beta x) \equiv \Gamma_r + j\Gamma_i \quad (\text{for } \alpha = 0)$$

where $r_0 = r(0)$ and $\chi_0 = \chi(0)$. In the complex Γ plane, the curves of constant Γ are circles [Figure 6-4(a)]. The curves of constant r are also curves [Figure 6-4(b)]. The curves of constant χ are circular arcs [Figure 6-4(c)]. Some important "corresponding" points on the Smith chart are shown in Table 6-1.

Table 6-1 Important Smith Chart Values

Condition	Γ	r	χ
Open-circuit	$1\angle 0°$	∞ (arbitrary)	arbitrary (∞)
Short-circuit	$1\angle 180°$	0	0
Pure reactance	$1\angle \pm 90°$	0	± 1
Matched line	0	1	0

The complete Smith chart of Figure 6-3 is obtained by superposing Figures 6-4(b) and (c). The circles of constant $|\Gamma|$ are not included; instead the value of $|\Gamma|$ corresponding to a point (r,χ) is read off of the left-hand external scale. The value of the VSWR is read from the right-hand scale. The two circumferential distance scales are in fractions of a wavelength.

Note!

From $(r,\chi) = (0,0)$, the outer scale goes clockwise *toward the generator* (i.e., measures x/λ), and the inner scale goes counterclockwise *toward the load* (i.e., measures d/λ).

Once around the chart is one-half wavelength. The third circumferential scale gives the reflection coefficient angle $\phi_\Gamma = \phi_R - 2\beta x$.

The chart can also be used for normalized admittances

$$y(x) \equiv Y(x)\, R_0 = g(x) + jb(x)$$

where r-circles are used for g, χ-arcs are used for b, the angle of Γ for a given y is $180^\circ + \phi_\Gamma$, and the point $y = 0 + j\,0$ is an open-circuit.

At high frequencies, it is essential to operate at minimum *VSWR* (ideally, at *VSWR* = 1). Several methods are used to match a load Z_R to the line, or to match cascaded lines with different characteristic impedances. Matching networks can be placed at the load ($x = 0$) or at some position $x = x_1$ along the line, as in Figure 6-5.

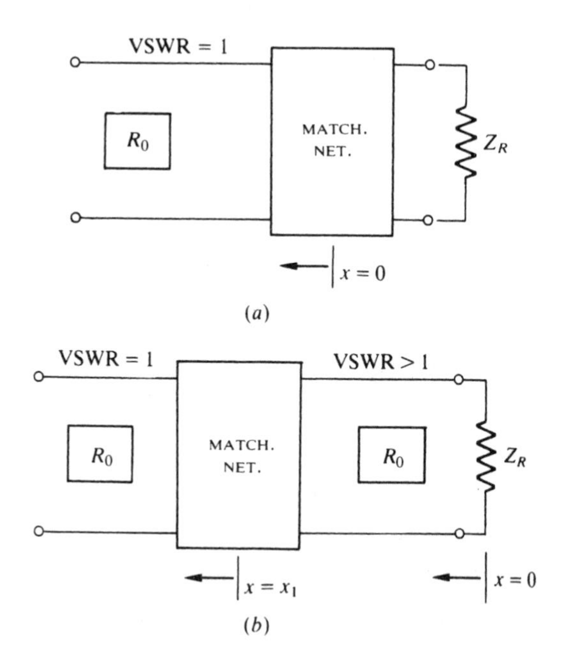

Figure 6-5 Matching network schemes.

The two sets of normalized conditions are as follows (refer to Figure 6-5):

(a) Before match: $z(0) = z_R = r_0 + jx_0; y(0) = g_0 + jb_0; VSWR > 1$
After match: $z(0) = 1 + j0; y(0) = 1 + j0; VSWR = 1$
(b) At load: $z(0) = r_0 + jx_0; y(0) = g_0 + jb_0; VSWR(0) > 1$
Before match: $z(x_1) = r_1 + jx_1; y(x_1) = g_1 + jb_1; VSWR = VSWR(0)$
After match: $z(x_1) = 1 + j0; y(x_1) = 1 + j0; VSWR = 1$

The matching networks at lower (radio) frequencies can be made with lumped low-loss reactive components; one lumped *L-C* network is shown in Figure 6-6. If Z_R has a reactive component, a reactance of opposite sign is added in series so that $Z'_R = R + j0$. Then, for a match,

$$Y_{in} = j\omega C_2 + \frac{1}{R + j\omega L_1} = \frac{1}{R_0}$$

or

$$L_1 = \frac{1}{\omega}\sqrt{R(R_0 - R)} \quad \text{and} \quad C_2 = \frac{L_1}{R\,R_0}$$

If $R > R_0$, the capacitor should be connected to the other end of the inductor.

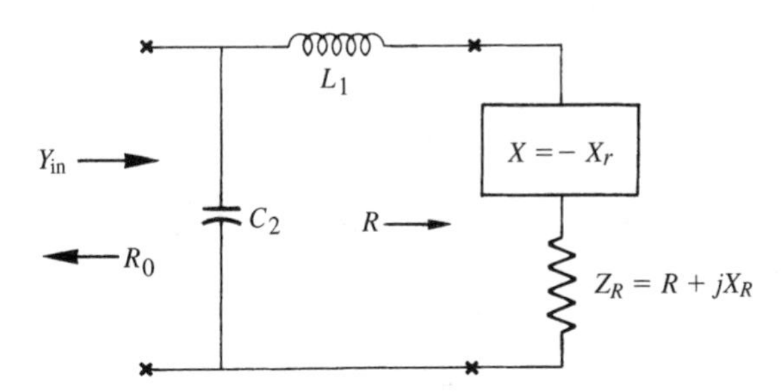

Figure 6-6 Tuning with reactive components.

To minimize dissipation losses at higher frequencies, a length of open- or short-circuited line is used for matching. One approach is to use a *single-stub* configuration. The configuration in Figure 6-7 uses one shorted stub.

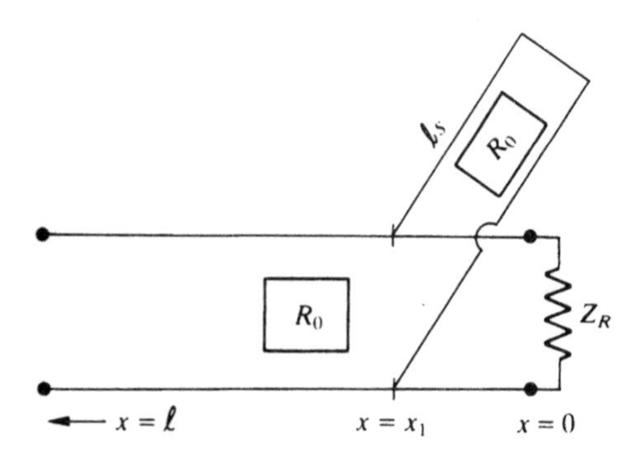

Figure 6-7 Single-stub matching.

To accomplish matching
(1) Determine x_1 such that $y(x_1) = 1 + jb_1$
(2) Determine ℓ_s such that $y(\ell_s) = 0 - jb_1$

After matching, $y(x_1) = 1 + j0$ and $VSWR = 1$ from x_1 to ℓ.

Example 6-1 The above two steps for single-stub matching may be accomplished using a Smith chart.

Solution: See Figure 6-8.

(i) Plot y_R and trace the $|\Gamma_R|$ [or $VSWR(0)$] circle.
(ii) Mark the intersections of the $|\Gamma_R|$ circle and the circle $g = 1$
(iii) From y_R move *toward the generator* to the first intersection, read $y_1 = 1 + jb_1$, and note the distance x_1 as a fraction of a wavelength.
(iv) Mark the point $y_1 = 0 - jb_1$ on the $|\Gamma|$ circle. From the short circuit position $y = \infty$, move *toward the generator* to the point $y = -jb_1$. Note the distance ℓ_s as a fraction of a wavelength. If the first intersection is not accessible, the second one can be used by readjusting the stub length for the susceptance at the new position.

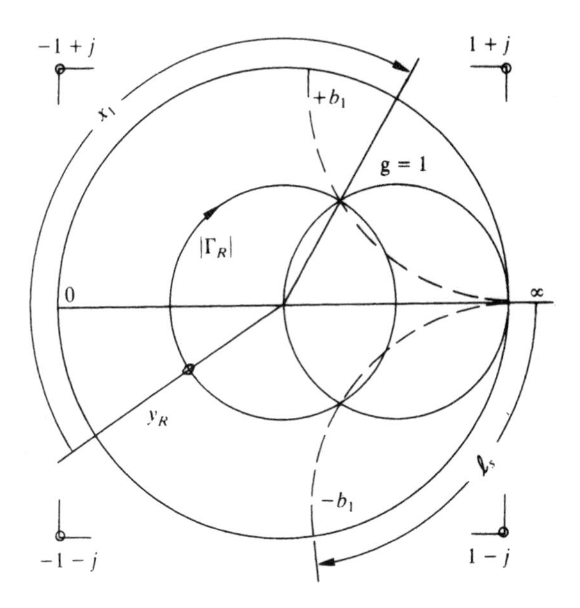

Figure 6-8 Using the Smith chart with single-stub matching.

A *slotted line* is used with high-frequency coaxial lines to measure *VSWR* and to locate voltage minima on the line. With the aid of the Smith chart, the impedance of an unknown termination can easily be found from the *VSWR* and the shift of a voltage minimum from a short-circuit reference position.

In Figure 6-9, the slotted line is inserted at a convenient terminal. With the Z_R in place, a probe is moved along the line to locate and measure maximum and minimum voltages. A suitable amplifier/indicator converts the probe output to a *VSWR* reading. Z_R is replaced by a short circuit and the reference minima are located for this high-*VSWR* condition. As would be expected, maxima and minima alternate at intervals of $\lambda/4$.

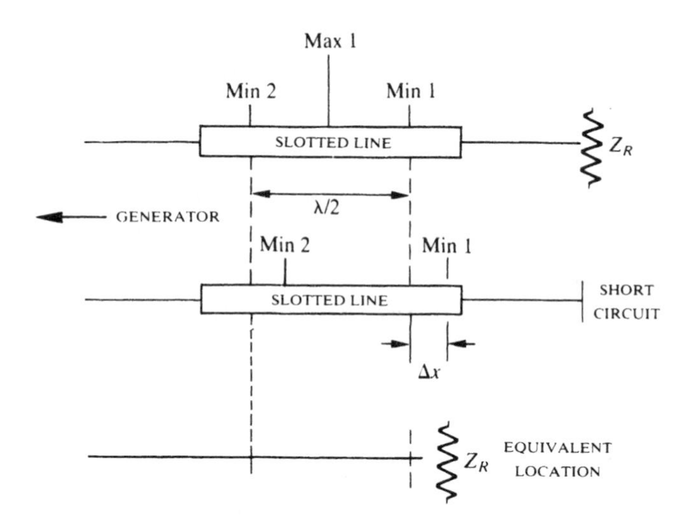

Figure 6-9 Using a slotted line to measure Z_R.

To find z_R with the Smith chart, draw the measured *VSWR* circle as in Figure 6-10 and locate the voltage minimum line (from 0 to 1 on the $\chi = 0$ line). Convert the measured Δx (See Figure 6-9) to wavelengths and mark the points on the *VSWR* circle that are Δx from the V_{min} line. The correct z_R is capacitive; a rotation through Δx *toward the generator* takes it into a V_{min} point. (If z_R were inductive, Δx would be greater than a quarter wavelength and a V_{max} point would occur before the V_{min} point.)

Transients in Lossless Lines

In switching applications and pulse operation, a change in voltage is suddenly applied to a section of line. An analysis of this transient condition generally requires recourse to time PDEs or to their Laplace transforms. However, in the special case of a lossless line (R = G = 0), $R_0 = \sqrt{L/C}$, velocity $u_p = 1/\sqrt{LC}$, a simple graphical method is available, based upon superposition of multiple reflected waves.

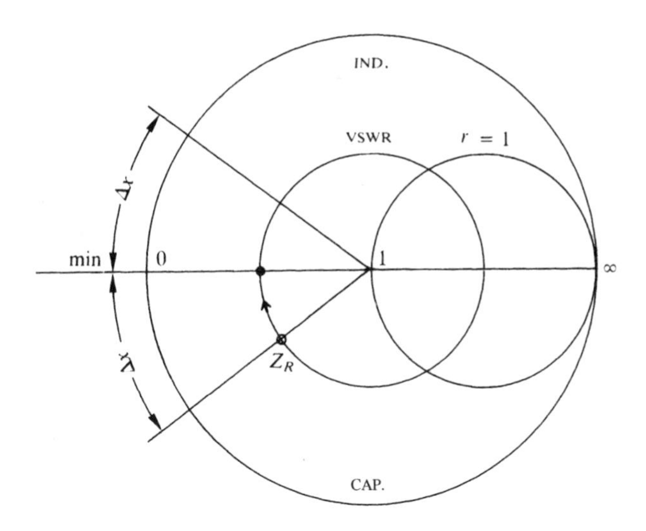

Figure 6-10 Determining unknown impedance.

Figure 6-11 shows a model for the lossless system, in which the exciting voltage $v_g(t)$ is switched on at $t = 0$ and where R_g is the source resistance.

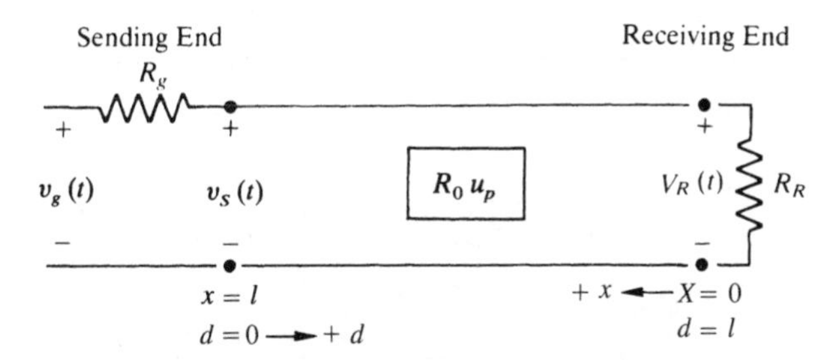

Figure 6-11 Lossless transmission line.

Now, an abrupt change at one end of the line has an effect at the other end only after one *delay time*, $t_D = l/u_p$, has elapsed. Reflection will occur at the receiving end if the load is not matched to the line ($R_R \neq R_0$). Similarly, if a wave is traveling in the $+x$-direction, a reflection will also occur at the sending end if the source is not matched ($R_g \neq R_0$).

Example 6-2 For the case where $v_g(t)$ is a 10-V step at $t = 0$ (i.e., a dc voltage) and where the line is matched at both ends ($R_R = R_g = R_0$), evaluate the transient voltage on the line.

Solution: The transient voltage conditions are displayed in *the time-distance plots* of Figure 6-12. Because the source is constant and three is no reflection at the receiving end, the system reaches a steady state, $v(d,t) =$ 5 V after only one delay time.

Example 6-3 Assume that everything is as in Example 6-2 with the exception of the load, which is now an open circuit ($R_R = \infty$). Evaluate the transient voltage on the line.

Solution: Figure 6-13 gives the time-distance plots. The reflection coefficient of the load is $\Gamma_R = 1$. Because of the single reflection at the load, a uniform steady state of 10 V is attained after $2t_D$.

Important Things to Remember:

- ✔ Transmission lines provide for guided propagation of energy.
- ✔ In transmission line analysis, the voltages and currents are functions of both position and time.
- ✔ Standing waves are set up if the reflection coefficient of the load is not zero.
- ✔ The Smith chart can be used to aid in the solution of high-frequency transmission line problems.
- ✔ In transient conditions, if an abrupt change happens at the input to a transmission line, there will not be a change at the load end until at least one delay time has occurred.

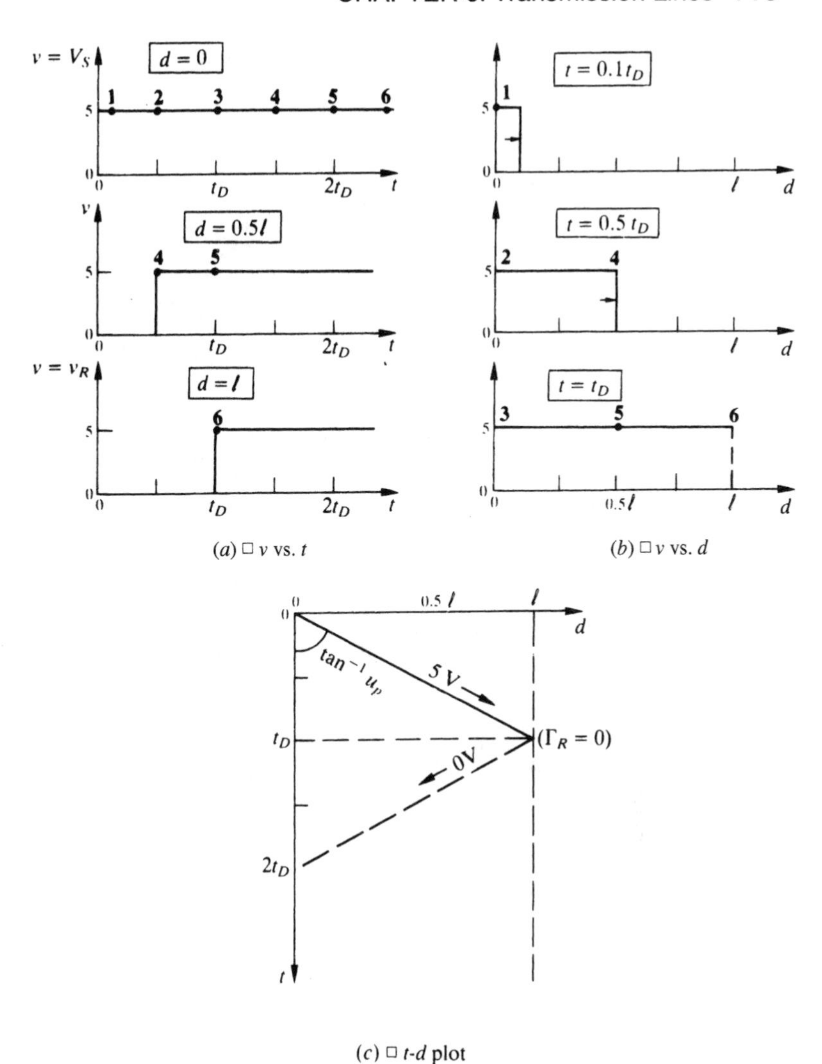

(*a*) □ v vs. t

(*b*) □ v vs. d

(*c*) □ t-d plot

Figure 6-12 Transient analysis with matched conditions.

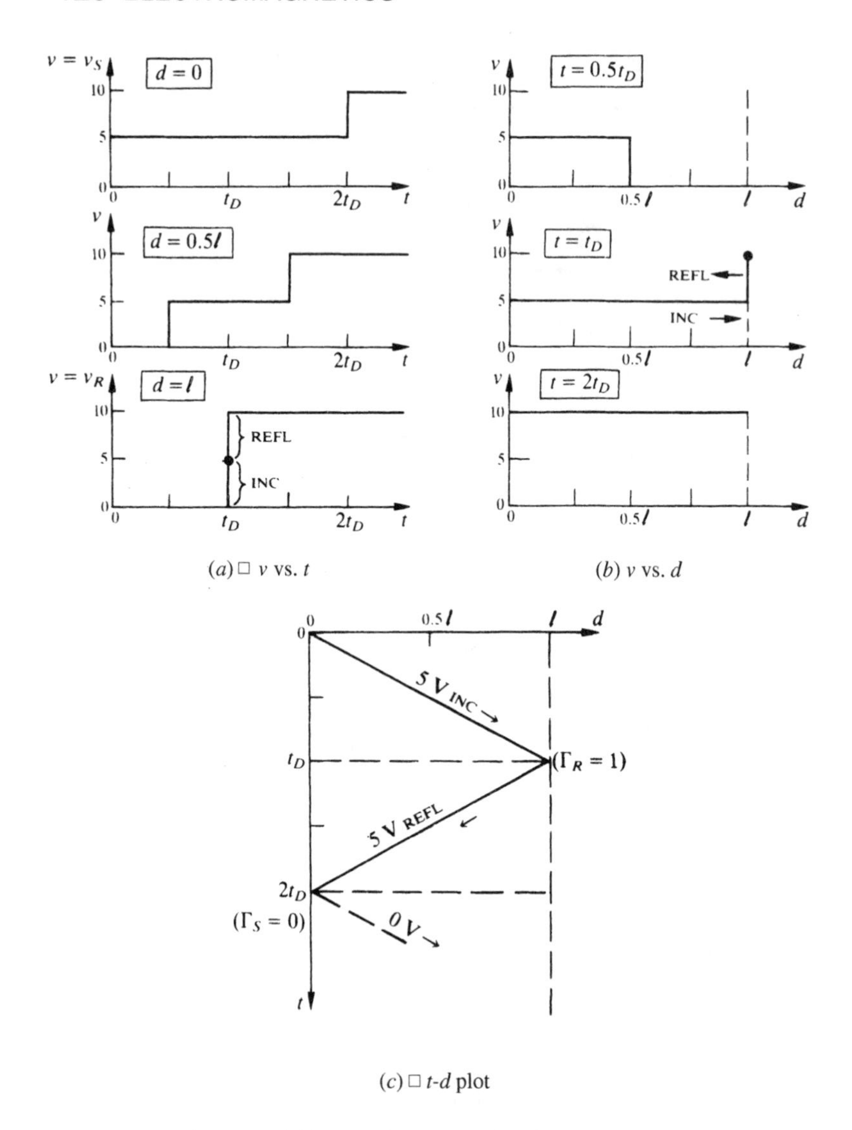

(*a*) □ *v* vs. *t*

(*b*) *v* vs. *d*

(*c*) □ *t*-*d* plot

Figure 6-13 Transient analysis with $(R_R = \infty)$.

Solved Problems

Solved Problem 6.1 A parallel-wire transmission line is constructed of #6 AWG copper wire (dia. = 0.162 in., $\sigma_c = 5.8\times10^7$ S/m) with a 12-inch separation in air. Neglecting internal inductance, find the per-meter values of *L, C, G,* the dc resistance and the ac resistance at 1 MHz.

Solution: The four geometrical factors for the parallel-wire transmission line involve the conductor radius, $a = 2.06\times10^{-3}$ m, and the separation d = 0.305 m. Since $d >> a$,

$$\text{GF}_\text{L} = \ln\left(\frac{d}{a}\right) = 5.0 \qquad \text{GF}_\text{C} = \frac{1}{\text{GF}_\text{L}} = 0.20$$

$$\text{GF}_\text{RDC} = \frac{2}{a^2} = 4.72\times10^5 \text{ m}^{-2} \qquad \text{GF}_\text{RAC} = \frac{2}{a} = 971 \text{ m}^{-1}$$

For air dielectric, $\mu_d = \mu_0$, $\varepsilon_d = \varepsilon_0$ and $\sigma_d = 0$. For copper, $\mu_c \approx \mu_0$ and the transmission line parameters are

$$L = \frac{\mu_d}{\pi}(\text{GF}_\text{L}) = 2.0\ \mu\text{H/m} \qquad R_{dc} = \frac{1}{\sigma_c \pi}(\text{GF}_\text{RDC}) = 2.59\times10^{-3}\ \Omega/\text{m}$$

$$C = \pi\varepsilon_d(\text{GF}_\text{C}) = 5.56 \text{ pF/m} \qquad \delta = \frac{1}{\sqrt{\pi f \mu_c \sigma_c}} = 66\ \mu\text{m}$$

$$G = 0 \text{ S/m} \qquad R_{ac} = \frac{1}{2\pi\sigma_c\delta}(\text{GF}_\text{RAC}) = 4.04\times10^{-2}\ \ \Omega/\text{m}$$

Solved Problem 6.2 For the parallel-wire line of problem ASP 6.1, find the characteristic impedance, propagation constant (attenuation and phase shift), velocity of propagation and wavelength for $f = 5$ kHz.

Solution: At 5 kHz, the dc resistance may be used.

$$\begin{aligned} Z &= R + j\omega L = 2.59\times10^{-3} + j(2\pi)(5000)(2\times10^{-6}) \\ &= 6.29\times10^{-2}\angle 87.6^\circ\ \ \Omega/\text{m} \end{aligned}$$

$$Y = G + j\omega C = j(2\pi)(5000)(5.56\times10^{-12}) = 1.747\times10^{-7}\angle 90^\circ\ \ \text{S/m}$$

$$Z_0 = \sqrt{\frac{Z}{Y}} = 600\angle -1.2° \ \Omega$$

$$\gamma = \sqrt{ZY} = 1.048 \times 10^{-4} \angle 88.8° = (2.19 \times 10^{-6}) + j(1.048 \times 10^{-4}) \ \text{m}^{-1}$$

Then

$$\alpha = 2.19 \times 10^{-6} \ \text{Np/m}, \beta = 1.048 \times 10^{-4} \ \text{rad/m},$$
$$u_p = \omega / \beta = 2.998 \times 10^8 \ \text{m/s}$$
$$\text{and } \lambda = 2\pi / \beta = 59.96 \ \text{km}.$$

Solved Problem 6.3 A 70-Ω transmission line is used at a frequency where $\lambda = 80$ cm with a load at $x = 0$ of $(140+j91)$ Ω. Use the Smith Chart to find Γ_R, *VSWR*, distance to the first voltage maximum from the load, distance to the first voltage minimum from the load, the impedance at V_{max}, the impedance at V_{min}, the input impedance for a section of the line that is 54 cm long, and the input admittance of that line with the load attached.

Solution: On the Smith chart, plot the normalized load

$$z_R = Z_R / R_0 = 2 + j1.3$$

as shown in Figure 6-14. Draw a radial line from the center through this point to the outer λ-circle. Read the angle of Γ_R on the angle scale: $\phi_R =$ 29°. Measure the distance from the center to the z-point and determine the magnitudes of Γ_R and *VSWR* from the scales at the bottom of the chart:

$$|\Gamma_R| = 0.50 \Rightarrow \Gamma_R = 0.5\angle 29° \quad \text{and} \quad VSWR = 3.0$$

Draw a circle at the center passing through the plotted normalized impedance. Note that this circle intersects the horizontal line at $3 + j0$. This point of intersection could be used to determine the *VSWR* instead of the bottom scale, because the circle represents a constant *VSWR*. Locate the intersection of the *VSWR* circle and the radial line from the center to the open-circuit point at the right of the z-chart. This intersection is the point where the voltage is a maximum (the current is a minimum) and the impedance is a maximum. The normalized impedance at this point is $3 + j0$, hence, $Z_{max} = 210 + j0 \ \Omega$. To find the distance from the load to the first V_{max}, use the outer scale (*wavelengths towards the generator*). The reference position is 0.21λ and the max. line is at 0.25λ. So, the distance is 0.04λ toward the generator or 3.2 cm from the load.

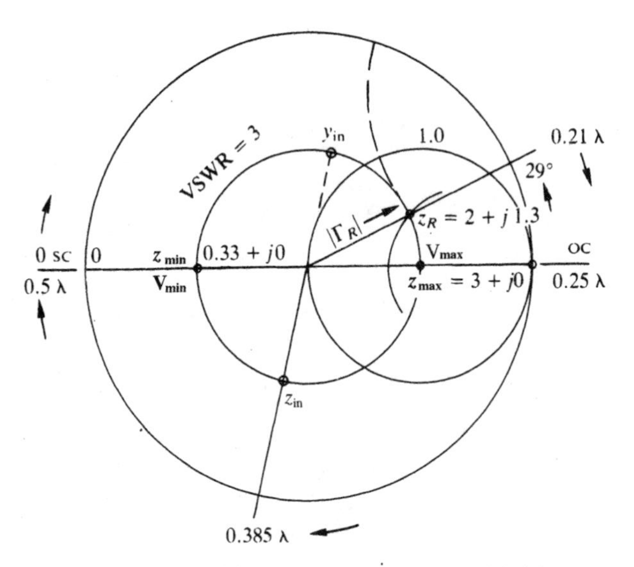

Figure 6-14 Smith chart for Problem ASP 6.3.

From the V_{max} point, move 0. 25λ toward the generator and locate the V_{min} point. The normalized impedance is 0.33 + j0 and Z_{min} = 23.1 + j0 Ω. The distance from the load to the first minimum is

$$0.25\lambda + 0.04\lambda = 0.29\lambda = 23.2 \text{ cm}$$

To find the input impedance of the 54 cm line, move $\frac{54}{80} = 0.675\lambda$ from the load toward the generator and read the normalized impedance. Once around the circle is 0.5λ, so locate the point that is 0.175λ from the load on the outer scale. The point is at $0.21\lambda + 0.175\lambda = 0.385\lambda$. Through this point draw a radial line and locate the intersection with the *VSWR* circle. The normalized input impedance is $z_{in} = 0.56 - j0.71$ and

$$Z_{in} = 39.2 - j\,49.7\ \Omega.$$

The normalized input admittance is located a diameter across on the chart, which corresponds to the inversion of a complex number. For z_{in} = 0.56 – j0.71, $y_{in} = 0.68 + j0.87$ and the input admittance is

$$Y_{in} = \frac{y_{in}}{R_0} = (9.71 + j12.4)\ \text{mS}$$

Chapter 7
ANTENNAS

IN THIS CHAPTER:

- ✔ *Magnetic Vector Potential and Radiated Fields*
- ✔ *Hertzian Dipole Antenna*
- ✔ *Antenna Parameters*
- ✔ *Finite-Length Dipole and Monopole*
- ✔ *Solved Problems*

Maxwell's equations as examined in Chapter 5 predict propagating plane waves in an unbounded source-free medium. In this chapter, the propagating waves produced by current sources are examined. The current sources are the models for *antennas*. In general, these waves have spherical wavefronts and direction-dependent amplitudes. Free-space conditions are exclusively assumed throughout the chapter.

Magnetic Vector Potential and Radiated Fields

In Chapter 2, the electric field intensity $\mathbf{E}$ was first obtained from known charge configurations. Later, the electric scalar potential V was developed and it was found that $\mathbf{E}$ could be obtained as a function of V:

$$\mathbf{E} = -\nabla V$$

Similarly, a *magnetic vector potential* **A**, defined such that

$$\mathbf{B} = \nabla \times \mathbf{A}$$

serves as an intermediate quantity from which **B,** and, hence, **H** can be calculated as well as **E**. Rewriting in terms of **H**,

$$\mathbf{H} = \frac{1}{\mu}\nabla \times \mathbf{A} = \frac{u}{\eta}\nabla \times \mathbf{A} \quad (1)$$

in which $u = 3\times 10^8$ m/s and $\eta = 120\pi\ \Omega$. Using Faraday's law (Chapter 4), one can find the electric field

$$\mathbf{E} = \frac{1}{j\omega\varepsilon}\nabla \times \mathbf{H} = \frac{1}{j\omega\mu\varepsilon}\nabla \times \nabla \times \mathbf{A} = \frac{u}{j\beta}\nabla \times \nabla \times \mathbf{A} \quad (2)$$

The phasor **A** is itself given by

$$\mathbf{A} = \int_{vol} \frac{\mu(\mathbf{J}_s e^{-j\beta R})}{4\pi R}\, dv \quad (3)$$

In (3), R is the distance between the observation point and the source current element $\mathbf{J}_s\, dv$. The significance of the factor $e^{-j\beta R}$ becomes clear when **A** is transformed to the time domain:

$$\mathbf{A} = \int_{vol} \frac{\mu \mathbf{J}_s \cos(\omega t - R/u)}{4\pi R}\, dv$$

Thus, **A** at the observation point properly reflects conditions at the source at earlier times—the lag for any given source element being precisely the time *R/u* needed for the condition to propagate to the observation point.

Hertzian Dipole Antenna

The magnetic vector potential set up by a Hertzian dipole (the infinitesimal current element of Figure 7-1) is, using (3),

$$\mathbf{A}(P) = \frac{\mu e^{-j\beta r}}{4\pi r}(I\, d\ell)\mathbf{a}_z$$

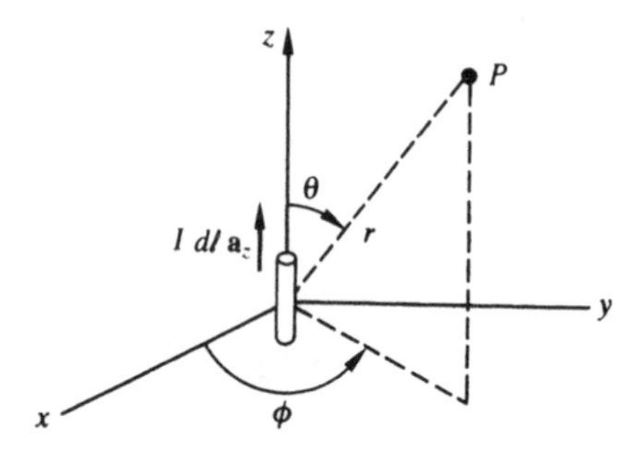

Figure 7-1 Hertzian dipole antenna.

In spherical coordinates, $\mathbf{a}_z = \cos\theta\,\mathbf{a}_r - \sin\theta\,\mathbf{a}_\theta$, and the magnetic and electric fields are, using (1) and (2),

$$H_\phi = \frac{I\,d\ell}{4\pi}\beta^2 \sin\theta\, e^{-j\beta r}\left[j\frac{1}{\beta r} + \frac{1}{\beta^2 r^2}\right] \tag{4}$$

$$E_r = \eta\frac{2I\,d\ell}{4\pi}\beta^2 \cos\theta\, e^{-j\beta r}\left[\frac{1}{\beta^2 r^2} - j\frac{1}{\beta^3 r^3}\right] \tag{5}$$

$$E_\theta = \eta\frac{I\,d\ell}{4\pi}\beta^2 \sin\theta\, e^{-j\beta r}\left[j\frac{1}{\beta r} + \frac{1}{\beta^2 r^2} - j\frac{1}{\beta^3 r^3}\right] \tag{6}$$

All other components are zero.

Note!

Attention will be restricted to the *far field* ($r >> 1$) in which terms on the order of $1/r^2$ and $1/r^3$ are neglected relative to the $1/r$-terms (i.e., for $r >> 1$, $1/r^2$ and $1/r^3$ are both $<< 1/r$).

Using (4), (5) and (6), the far fields are

$$H_\phi = j\frac{I\,d\ell\beta}{4\pi r}\sin\theta\, e^{-j\beta r} \tag{7}$$

$$E_\theta = \eta\, j\frac{I\,d\ell\beta}{4\pi r}\sin\theta\, e^{-j\beta r} = \eta\, H_\phi \tag{8}$$

It is clear that (7) and (8) represent diverging spherical waves which at any point are traveling in the $+\mathbf{a}_r$-direction with an amplitude that falls off as $1/r$.

The power radiated by the antenna is obtained from the *time-average Poynting vector* . The Poynting vector is the power flux density of the radiated field and has units of W/m^2. The average Poynting vector is defined as

$$\mathscr{P}_{avg} = \tfrac{1}{2}\mathrm{Re}(\mathbf{E}\times\mathbf{H}^*) \tag{9}$$

where $\mathbf{H}^*$ is the complex conjugate of $\mathbf{H}$. This follows the power calculation in circuit analysis for which the average power is $P_{avg} = \tfrac{1}{2}\mathrm{Re}(VI^*)$. To find the average power radiated by the Hertzian dipole, integrate $\mathscr{P}_{avg}$ over the surface of a sphere that encloses the antenna:

$$\begin{aligned} P_{rad} &= \int_0^{2\pi}\int_0^{\pi}\mathscr{P}_{avg}\bullet d\mathbf{S}_{sphere} = \int_0^{2\pi}\int_0^{\pi}\mathscr{P}_{avg}\bullet(r^2\sin\theta\, d\theta\, d\phi\, \mathbf{a}_r) \\ &= \int_0^{2\pi}\int_0^{\pi}\left[\tfrac{1}{2}\mathrm{Re}(E_\theta H_\phi^*)\right]r^2\sin\theta\, d\theta\, d\phi \\ &= \frac{\eta(\beta I\, d\ell)^2}{12\pi} = \frac{\eta\pi I^2}{3}\left(\frac{d\ell}{\lambda}\right)^2 \quad \text{(W)} \end{aligned} \tag{10}$$

Antenna Parameters

The *radiation resistance* R_{rad} is defined as the value of a hypothetical resistor that would dissipate a power equal to the power radiated by the antenna when fed by the same current, thus $P_{rad} = \tfrac{1}{2}I_0^2 R_{rad}$, $R_{rad} = 2P_{rad}/I_0^2$ where I_0 is the peak value of the feed point current. For the Hertzian dipole, using (10),

$$R_{rad} = \frac{2\pi\eta}{3}\left(\frac{d\ell}{\lambda}\right)^2 \approx 790\left(\frac{d\ell}{\lambda}\right)^2 \quad (\Omega)$$

The *radiation pattern* $F(\theta,\phi)$ gives the relative variation of the far-field **E** or **H** magnitude versus direction. The distance r is assumed to be a constant. For the Hertzian dipole, this reduces to $F(\theta) = \sin\theta$, since $|\mathbf{E}|$ and $|\mathbf{H}|$ are independent of ϕ.

The *radiation intensity* $U(\theta,\phi)$ is another measure of antenna performance. It is defined as the time-average radiated power per unit solid angle—i.e., it is a function of power versus angle (see Figure 7-2). *Be aware that the symbol for solid angle is also* Ω, *not to be confused with the units of resistance, ohms.* Using the expression for the average power (10),

$$P_{rad} = \int_0^{2\pi}\int_0^{\pi} \mathscr{P}_{avg} \bullet d\mathbf{S}_{sphere} = \int_0^{2\pi}\int_0^{\pi} (\mathscr{P}_{avg} \bullet r^2\mathbf{a}_r)\sin\theta \, d\theta \, d\phi$$
$$= \int_0^{2\pi}\int_0^{\pi} U(\theta,\phi)\sin\theta \, d\theta \, d\phi = \int_0^{2\pi}\int_0^{\pi} U(\theta,\phi) \, d\Omega$$

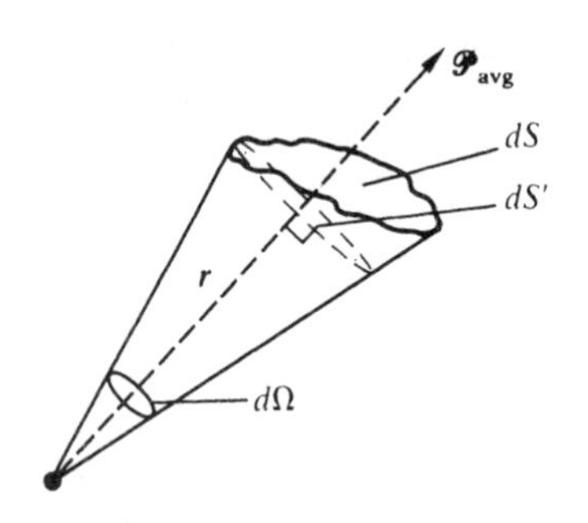

Figure 7-2 Defining the radiation intensity $U(\theta,\phi)$.

Therefore, the radiation intensity and element of solid angle are

$$\Rightarrow U(\theta,\phi) \equiv (\mathscr{P}_{avg} \bullet r^2\mathbf{a}_r) \quad (\mathrm{W/sr})$$
$$\Rightarrow d\Omega \equiv \sin\theta \, d\theta \, d\phi \quad (\mathrm{sr})$$

where sr is a *steradian.* Because U is independent of r (by energy conservation), the far field may be used in its evaluation. For the Hertzian dipole, using (7) through (10),

$$U(\theta) = \frac{\eta}{8}\left(\frac{I\,d\ell}{\lambda}\right)^2 \sin^2\theta \quad (\mathrm{W/sr}) \tag{11}$$

Polar plots of the radiation pattern $F(\theta)$ and radiation intensity $U(\theta)$ for the Hertzian dipole are given in Figure 7-3.

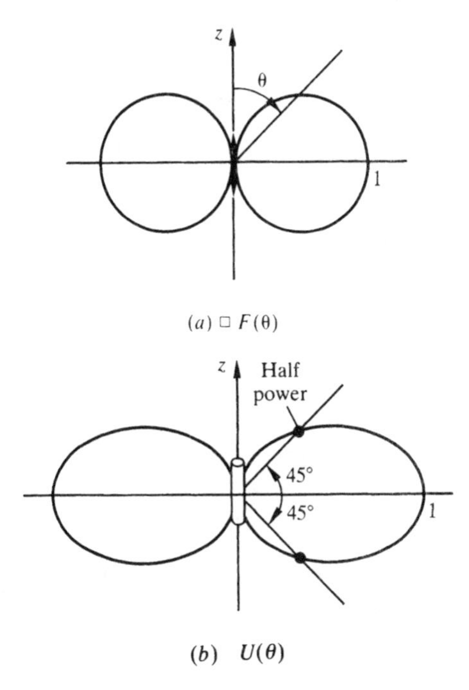

Figure 7-3 Radiation pattern *F* and radiation intensity *U*.

In Figure 7-3(*b*), the *half-power* points are at $\theta_{HP1} = 45°$ and $\theta_{HP2} = 135°$. The *half-power beam width* θ_{HP} is therefore $\theta_{HP} = \theta_{HP2} - \theta_{HP1} = 90°$.

You Need to Know 

In general, the smaller the beam width (about the direction of U_{max}), the more *directive* the antenna —i.e., the radiated power is focused over a smaller angular sector.

Directive gain $D(\theta,\phi)$ of an antenna is defined as the ratio of the radiation intensity $U(\theta,\phi)$ to that radiation intensity U_0 of a hypothetical *isotropic* radiator that radiates the same total power P_{rad} as the actual antenna. For the isotropic radiator,

$$U_0 = \frac{P_{rad}}{4\pi}$$

Then the directive gain of the antenna is then

$$D(\theta,\phi) = \frac{U(\theta,\phi)}{U_0} = \frac{4\pi U(\theta,\phi)}{P_{rad}} \tag{12}$$

The *directivity* D of the antenna is the *maximum* value of its directive gain

$$D = \frac{4\pi U_{max}}{P_{rad}} \tag{13}$$

For the Hertzian dipole, using (10) and (11) in (12) and (13) gives the directive gain

$$D(\theta,\phi) = \frac{(4\pi)\frac{\eta}{8}\left(\frac{I\,d\ell}{\lambda}\right)^2 \sin^2\theta}{\left(\frac{\eta\pi}{3}\right)\left(\frac{I\,d\ell}{\lambda}\right)^2} = 1.5\sin^2\theta$$

and directivity $D = 1.5$.

The *radiation efficiency* of an antenna is $e_{rad} \equiv P_{rad} / P_{in}$, where P_{in} is the time-average power that the antenna accepts from the feed. The *power gain* $G(\theta,\phi)$ is defined as the efficiency times the directive gain:

$$G(\theta,\phi) \equiv e_{rad}\, D(\theta,\phi) = \frac{4\pi U(\theta,\phi)}{P_{in}} = \frac{4\pi U(\theta,\phi)}{P_{rad} + P_L}$$

where, for wire antennas, P_L is primarily the ohmic power loss of the antenna. A lossless isotropic radiator has a power gain $G_0 = 1$. At times, the power gain of the antenna is expressed in decibels where

$$G_{dB} = 10\log_{10} G(\theta,\phi)$$

Finite-Length Dipole and Monopole

The expression (10) for the radiated power of the Hertzian dipole contains the term $(d\ell / \lambda)^2$ which suggests that the length should be comparable to the wavelength. The open-circuited two-wire transmission line shown in Figure 7-4(a) has currents on the conductors that are out of phase so that the far field nearly cancels out. An efficient antenna results when the line is opened out as shown in Figure 7-4(b), producing current phasors

$$I_1(z') = I_m \sin\beta\left(\frac{L}{2} - z'\right), \quad 0 < z' < L/2$$

and

$$I_2(z') = I_m \sin\beta\left(\frac{L}{2} + z'\right), \quad -L/2 < z' < 0$$

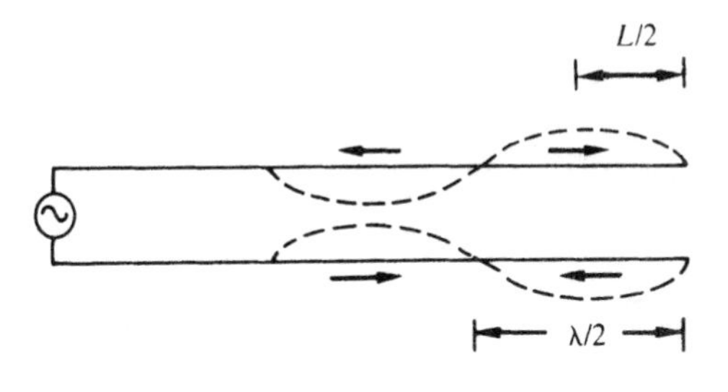

(a)

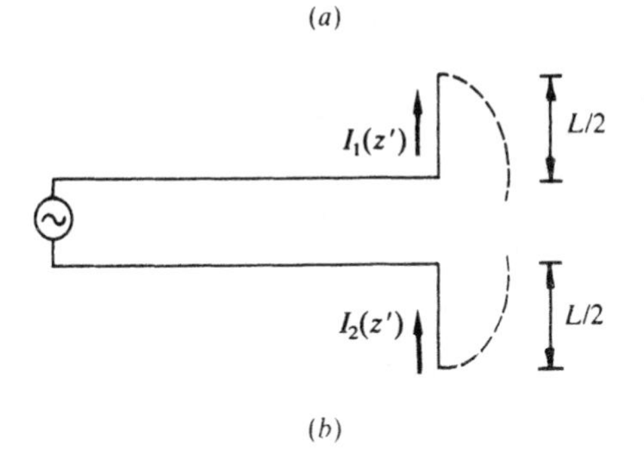

(b)

Figure 7-4 (a) Open-circuited transmission line and (b) dipole antenna.

The two currents are exactly in-phase at mirror-image points along the z-axis and they vanish at the endpoints $z' = \pm L/2$. The two legs form a single dipole antenna of finite length L. Note that the current at the feed point $z' = 0$ is related to the maximum current by

$$I_0 = I_m \sin\frac{\beta L}{2}$$

The far field is calculated by means of (1), (2), and (3), under the assumption that $r >> L$ and $r >> \lambda$.

$$H_\phi = \frac{jI_m e^{-j\beta r}}{2\pi r} F(\theta), \qquad E_\theta = \eta\, H_\phi$$

where the radiation pattern function is given by

$$F(\theta) = \frac{\cos\left(\beta\frac{L}{2}\cos\theta\right) - \cos\left(\beta\frac{L}{2}\right)}{\sin\theta}$$

For L up to about 1.2λ the antenna radiation patterns resemble the figure eight, becoming sharper as L approaches 1.2λ. In the other limit, as $L << \lambda$, the pattern is that of the Hertzian dipole shown in Figure 7-3(a). As L becomes greater than 1.2λ the patterns become multi-lobed (Figure 7-5).

The radiation resistance of a finite dipole of length $(2n - 1)\lambda/2$, n = 1,2,3, … , can be shown to be $R_{rad} = 30\ \text{Cin}[(4n - 2)\pi](\Omega)$ where

$$\text{Cin}(x) = \int_0^x \frac{1-\cos y}{y}\, dy$$

is a tabulated function.

For the important and widely-used *half-wave dipole* ($n = 1$), R_{rad} = 30(2.438)=73 Ω and $D = 1.64$.

A conductor of length $L/2$ normal to an infinite conducting ground plane [Figure 7-6(a)] forms a monopole antenna. *Image theory* can be used to form the equivalent problem of a dipole in free space [(see Figure 7-6(b)].

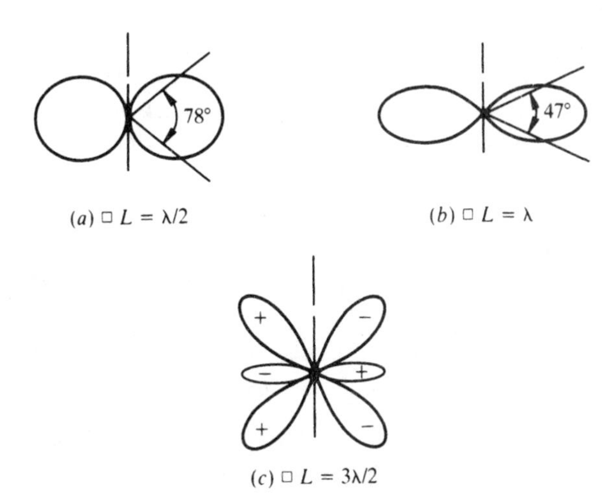

Figure 7-5 Radiation patterns for dipole antennas.

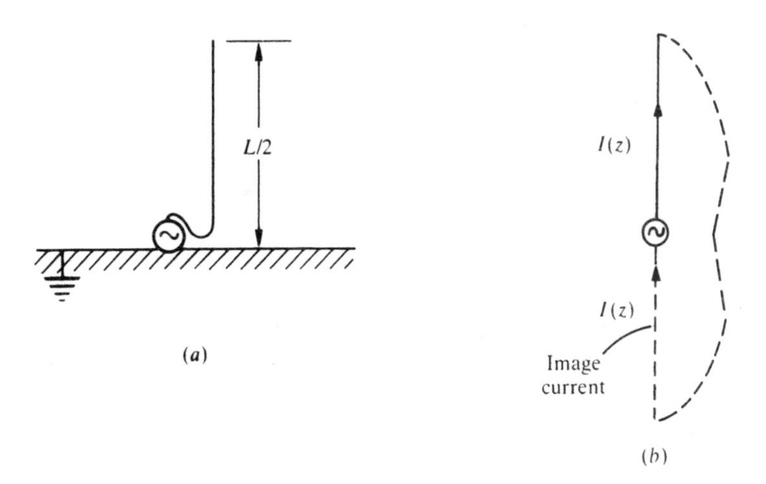

Figure 7-6 (*a*) Monopole antenna; (*b*) equivalent dipole antenna problem.

When fed at the base, the resulting **E** and **H** fields of the monopole antenna are identical to the dipole antenna above the location for the ground plane.

As the monopole radiates power only in the region above the ground plane, the total radiated power is one-half that of the corresponding dipole. From $R_{rad} = 2P_{rad}/I_0^2$, it follows that the radiation resistance of the monopole is one-half the value for the dipole. Thus, for $L/2 = \lambda/4$ (*quarter-wave monopole*), $R_{rad} = 36.5\ \Omega$.

Important Things to Remember:

- ✔ Antennas radiate diverging spherical waves.
- ✔ The directivity of the antenna is the maximum increase in radiation intensity *U* over that of an isotropic radiator transmitting the same radiated power.
- ✔ The dominant current component for a straight wire dipole is a sinusoidal standing wave.
- ✔ The monopole radiated fields are equal to the corresponding dipole fields above the ground plane.

Solved Problems

Solved Problem 7.1 A center-fed dipole with a *z*-directed current has electrical length $L/\lambda << 1/30$. (*a*) Show that the current distribution may be assumed to be triangular in form. (*b*) Find the components of the vector magnetic potential **A**.

Solution: (a) Since

$$\beta\left(\frac{L}{2} - |z'|\right) < \beta\frac{L}{2} = \pi\frac{L}{\lambda} << \frac{1}{10}$$

we have

$$I(z') = I_m \sin\beta\left(\frac{L}{2} - |z'|\right)$$

$$\approx I_m\beta\left(\frac{L}{2} - |z'|\right) \approx \frac{2I'_m}{L}\left(\frac{L}{2} - |z'|\right), \text{ where } I'_m = I_m\frac{\beta L}{2}$$

which is triangular in form.

(b) $$\mathbf{A} = \frac{\mu}{4\pi}\int_{-L/2}^{L/2} I(z')\mathbf{a}_z\left(\frac{e^{-j\beta r}}{4\pi r}\right)dz'$$

$$= \frac{2\mu I'_m e^{-j\beta r}}{4\pi L r}\int_{-L/2}^{L/2}\left(\frac{L}{2} - |z'|\right)\mathbf{a}_z\, dz' = \frac{\mu I'_m}{4\pi r}\left(\frac{L}{2}\right)e^{-j\beta r}\,\mathbf{a}_z$$

from which

$$\mathbf{A}_r = A_z\cos\theta = \frac{\mu I'_m}{4\pi r}\left(\frac{L}{2}\right)e^{-j\beta r}\cos\theta$$

$$\mathbf{A}_\theta = -A_z\sin\theta = -\frac{\mu I'_m}{4\pi r}\left(\frac{L}{2}\right)e^{-j\beta r}\sin\theta$$

$$\mathbf{A}_\phi = 0$$

Solved Problem 7.2 A Hertzian dipole of length $L = 2$ m operates at 1 MHz. Find the radiation efficiency if the copper conductor has $\sigma_c = 5.7\times10^7$ S/m, $\mu_r = 1$ and radius $a = 1$ mm.

Solution: The radiation efficiency is

$$e_{rad} = \frac{P_{rad}}{P_{in}} = \frac{P_{rad}}{P_{rad} + P_{loss}} = \frac{R_{rad}}{R_{rad} + R_{loss}}$$

where R_{rad} is the radiation resistance and R_{loss} is the ohmic resistance. The radius a is much greater than the skin depth

$$\delta = \frac{1}{\sqrt{\pi f\mu\sigma_c}} = \frac{1}{15}\text{ mm}$$

so that the current may be assumed to be confined to a cylindrical shell of thickness δ, giving an ac resistive loss of (similar to the parallel-wire parameters from Chapter 6)

$$R_{loss} = \frac{1}{\sigma_c} \frac{L}{(2\pi a)\delta} = 0.084\ \Omega$$

The radiation resistance is given by

$$R_{rad} = (790)\left(\frac{L}{\lambda}\right)^2 = (790)\left(\frac{Lf}{u}\right)^2 = 0.035\ \Omega$$

Therefore, the efficiency is

$$e_{rad} = \frac{0.035}{0.119} = 29.4\%$$

Solved Problem 7.3 (*a*) Find the current required to radiate a power of 100 W at 100 MHz from a 0.01-m Hertzian dipole. (*b*) Find the magnitudes of **E** and **H** at (100m, 90°, 0°).

Solution: (*a*) The wavelength is

$$\lambda = \frac{3\times 10^8}{10^8} = 3\text{ m}$$

For the radiation resistance,

$$R_{rad} = (790)\left(\frac{d\ell}{\lambda}\right)^2 = 8.78\times 10^{-3}\ \Omega = \frac{2P_{rad}}{I^2}$$

Solving for the current,

$$I = \sqrt{\frac{200}{8.78\times 10^{-3}}} = 151\text{ A}$$

(This extremely high current illustrates that an antenna with a length much less than a wavelength is not an efficient radiator.)

(*b*) $|\mathbf{E}| = \dfrac{\eta\beta I\, d\ell}{4\pi r}\sin 90° = 0.95\text{ V / m}$

$$|\mathbf{H}| = \frac{\beta I\, d\ell}{4\pi r}\sin 90° = 2.52\times 10^{-3}\text{ A / m}$$

Index